*A Textbook of*

# Aquaculture and Fisheries Resource

## THE AUTHORS

**Shri Ram Yadav** did his M.F.Sc. (Fisheries Resource Management) from CCS HAU, Hisar (Haryana). He has published several research publications on the fisheries sector in reputed journals. He is a member of ISAHRD, Chandigarh. His field of interest is aquaculture and fish disease management.

**Ambrish Singh** is pursuing his doctoral degree in Fisheries Resource at the College of Fisheries, ANDUA&T, Kumarganj, Ayodhya, Uttar Pradesh. Previously, he worked as a teaching faculty member at the College of Fisheries Science and Research Centre, Etawah (CSAUA&T), Kanpur, U.P. He did his M.F.Sc. (Fisheries Resource Management) from CCS HAU, Hisar (Haryana). Gold Medalist during B.F.Sc. (ANDUAT Kumargnj, Ayodhya, Uttar Pradesh). He qualified for ARSB NET in 2023. His field of interest is fisheries taxonomy, plankton diversity, fish diversity and disease management.

**Siddharth Kumar Jatav** is pursuing his doctoral degree in fisheries resource management at the College of Fisheries Science, CCS HAU Hisar (Haryana). He did his M.F.Sc. with NTS from Kamdhenu University, Veraval, Gujarat. He qualified for ARSB NET in 2023. He has been awarded the Young Scientist award from the Green Planet Agrotech Foundation and has been a life member of the Green Planet Agrotech Foundation. He has published several research publications on the fisheries sector in reputed journals. His field of interest is fish diversity and taxonomy, carp culture and breeding and shrimp culture.

**Dr. Anindya Sundar Barman** is an Associate Professor at Central Agricultural University (Imphal). He has more than 15 years of experience and is actively involved in many academic, research, and extension programmes in Fish Genetics and Biotechnology. He has published several research papers on the fisheries sector in reputed journals.

**Dr. Ravikant** works as an Assistant Professor (Stage-III) in the Department of Zoology and Associated faculty in the Department of Fisheries Resource Management and Aquaculture, College of Fisheries Science, CCS HAU, Hisar (Haryana). He has over 12 years of experience in teaching, research and extension. He was interested in building entrepreneurship in shrimp and fish cultures and involved himself in various research activities. He has authored four books and over a hundred research papers/ articles.

# *A Textbook of*
# Aquaculture and Fisheries Resource


Shri Ram Yadav

Ambrish Singh

Siddharth Kumar Jatav

Anindya Sundar Barman

Ravikant



ASTRAL

*Publishing Authors Across the Globe*

2025

Daya Publishing House®

*A Division of*

Astral International Pvt. Ltd.

New Delhi - 110002

ISBN: 978-93-5461-957-1 (HB)



*Published by* : **Daya Publishing House®**
*A Division of*
**Astral International Pvt. Ltd.**
– ISO 9001:2015 Certified Company –
4736/23, Ansari Road, Darya Ganj
New Delhi-110 002
Ph. 011-43549197, 8130496929
E-mail: info@astralint.com
Website: www.astralint.com

*Printed at* : **Replika Press Pvt Ltd**

*Dedicated to Our Mentors*

***Dr. S. K. Garg***

***&***

***Dr. R.K. Gupta***

# Preface

Indian fisheries occupy the second position in global fish production, with an annual growth rate of 4.7%. They record 3.2% growth in the marine sector and 6.2% in the inland sector, contributing 1.10% to the total GDP and 5.3% of the nation's agricultural GDP.

Fisheries and Aquaculture continue to serve as a vital source of sustenance, nourishment, income, and livelihood for millions of individuals. Fish is an affordable and valuable source of animal protein, making it a practical option for combating hunger and malnutrition within the country. Fish is a significant source of protein and other essential nutrients in the diet. Moreover, fisheries resource management has become a favourable subject for UG and PG students. Most agricultural universities have established Fisheries Resource management as separate departments in various fisheries colleges/ institutes/Universities. This book will be helpful not only for undergraduate and postgraduate students of agricultural universities but also for fish farmers and multiple stakeholders to achieve their objectives. This book will cover almost all the aspects of fisheries resource management. An effort has been made to cover various aspects of Fisheries management such as the Status of Indian Fisheries, Cultivation System, Site selection of fish farms, Pond Management, Biology of Indian major carps and exotic carps, Commercially important fish culture, Fish food organisms, Reservoir fisheries in India, Types of lakes, Prominent lakes of India Subcontinent, Riverine fisheries in India, Riverine fisheries in India, Traditional fishing crafts and gears used in India, Impact of environmental factors on inland fisheries, Impact of ecological factors on inland fisheries, Status of ornamental fisheries in India etc. It would also be helpful for students for the preparation of various competitive examinations like JRF, SRF, NET, ARS, AO (Agriculture Officer), UG and PG entrance.

We want to thank all the distinguished authors and researchers whose work and contributions have formed the basis of some key material for this book. The books and the literature we consulted are mentioned at the end of each chapter. We are highly thankful to our universities (CCS HAU, Hisar and Central Agricultural University, Imphal). for providing a platform for creating the manuscript. Last, we sincerely thanks to faculty members, Department of Zoology, CCS Haryana Agricultural University, Hisar, for their valuable suggestions and kind support during this book's preparation. The interest taken by the publisher in this book is also highly appreciated.

**S.R. Yadav**

**A. Singh**

**S. K. Jatav**

**A.S. Barman**

**Ravikant**

# Contents

CHAPTER

# Status of Indian Fisheries

1

## Introduction

Fisheries and Aquaculture continue to serve as a vital source of sustenance, nourishment, income, and livelihood for millions of individuals. Fish is an affordable and valuable source of animal protein, making it a practical option for combating hunger and malnutrition within the country. Fish is a significant source of protein and other essential nutrients in the diet. There are multiple methods for preparing fish, such as consuming it raw (as in sashimi), marinating (as in ceviche), pickling (as in pickled herring), or smoking (as in smoked salmon). Fish can also be cooked through baking, frying (as in fish and chips), grilling, poaching, or steaming. Although many preservation techniques have become obsolete in different cultures, they are still practised due to the distinct flavours and textures they impart to the consumed fish. Handling fish catch involves various processes, including marketing fresh fish, freezing, curing, canning, and reduction. Among these, new fish marketing is the most widely practised activity.

India has experienced consistent growth in the disposition of fish caught between 2013 and 2021. In 2020-21, Andhra Pradesh, West Bengal, and Odisha exhibited the highest levels of fish catch disposition. The Indian Fisheries sector has gradually developed over the years and has become a significant socio-economic aspect of the nation. The fisheries sector contributed approximately 1.1% to the Indian economy, while the agriculture sector accounted for about 6.72% in 2021-22. The fishing industry is vital to the country's economy and makes a considerable contribution to its foreign exchange earnings. India is among the top five fish-exporting nations globally, ranking third after China and Indonesia. India achieved record-breaking seafood exports, reaching 1.36 million metric tons valued at US$ 7.76 billion. India exported marine products to 123 countries, with the United States and China serving as the primary importers of Indian seafood. India's primary marine exports encompass an array of seafood commodities, comprising frozen shrimps, various fish varieties, cuttlefish, squids, dried seafood items, as well as live and chilled seafood products. The foremost player in this sector is frozen shrimp, both in terms of quantity, contributing to a substantial 53.18% of the total export volume, and export value, accounting for approximately 75.11% of the overall revenue generated. Frozen fish, cuttlefish, and squid, while relatively smaller contributors, still hold their significance within the market. Frozen fish represented 6% of the total export value of marine products in the fiscal year 2021-22. Cuttlefish and squid, each accounting for 3% and 5% of the overall export value, respectively, further diversify the marine product export portfolio from India.

## Current Status and Present Trend in Production

India contributes over 16% of the worlds total inland fish output and 5% of the world's marine fish production, making it the third-largest producer of fish and aquaculture products globally. In the financial year 2021-22, India recorded a total fish production of 162.48 lakh tonnes, comprising 121.21 lakh tonnes from inland sources and 41.27 lakh tonnes from marine sources.

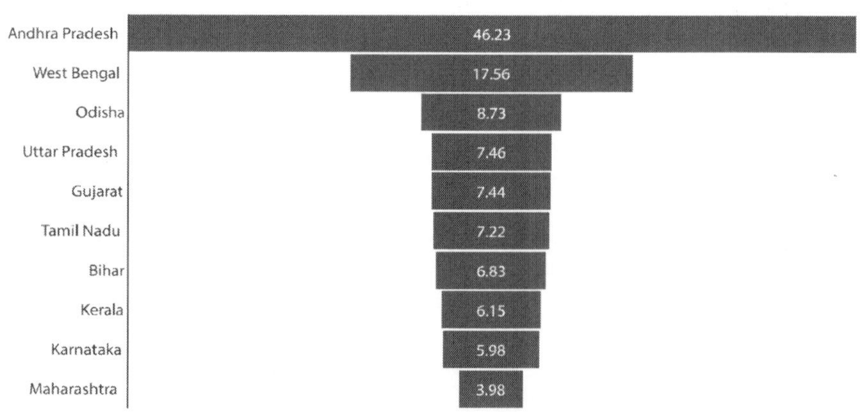


**Fig. Top most producing state of fish catch 2020-2021**

*(Source: Handbook on Fisheries Statistics 2022)*

Fish production in India has seen a remarkable increase from 56.56 lakh tonnes in 2000-01 to 162.48 lakh tonnes in 2021-22. Among the various states, Andhra Pradesh, West Bengal, Karnataka, Odisha, and Gujarat emerged as the five primary fish-producing states from 2021-22. India's annual fish production through capture and culture fisheries currently stands at 16.25 million tonnes, providing an affordable protein source that contributes to the nutritional security of the population. Additionally, exporting fish and fishery products annually contributes to the country's foreign exchange earnings.

States like Uttar Pradesh, Odisha, Bihar, Kerala, and Tamil Nadu exhibit the highest levels of total fish consumption. Additionally, the per capita consumption of fish and prawns is highest in Lakshadweep and Kerala, respectively. Tripura has the highest number of households (both rural and urban) that consume fish.

## Status of World Fish Production

In 2020, the worldwide production of fisheries and aquaculture achieved a significant milestone, surging to a grand total of 214 million tons. This impressive figure encompassed a substantial 36 million tons of algae and an even more substantial 178 million tons of aquatic organisms. Comparing this accomplishment to the previous record set in 2018, which was 213 million tonnes, there was a little 3 percent increase. However, the significant drop of 4.4 percent in the harvest of marine resources had a substantial dampening effect on this progress. This fall was influenced by decreased catches of pelagic species, notably Anchoveta, as well as decreased catches in China and the COVID-19 pandemic's effects in 2020. Even while the growth rate in aquaculture has slowed over the previous two years, it continues to increase, somewhat offsetting the decline in catch fisheries. The overall pattern in the production of aquatic creatures conceals notable variations among different continents, regions, and countries. Asia, responsible for a substantial 70% share of worldwide production in 2020, assumed the top position as the predominant producer, with the Americas, Europe, Africa, and Oceania following in that order. With a 35 percent share of global output, China maintained its position as the top producer. The rise of aquatic animal production in inland waterways has been greatly aided by the growth of aquaculture during the past several decades. In the late 1980s, inland waters contributed merely 12 percent to the total output, whereas by 2020, this figure had surged to 37 percent.

In 2020, the collective output of capture fisheries, excluding algae, amounted to 90.3 million tonnes, carrying an approximate monetary worth of USD 141 billion. This extensive yield comprised 11.5 million tonnes sourced from inland water bodies and a substantial 78.8 million tonnes extracted from marine environments. In comparison to the average of the preceding three years, this production figure signaled a 4.0 percent decrease. Remarkably, anchoveta continued to dominate as the most prevalent species captured, constituting more than 85% of the entire marine catch production. Furthermore, the four most esteemed catch categories—namely tunas, cephalopods, prawns, and lobsters—remained relatively stable at their elevated levels throughout 2020, experiencing only marginal declines, if any.

The worldwide catches of freshwater fish, which have been calculated to be around 11.5 million metric tonnes, experienced a decline of 5.1 percent from the year 2019. Nevertheless, these figures still stand at historically elevated levels, primarily due to improved data reporting from the countries responsible for the production. The continents of Africa and Asia jointly contributed to approximately two-thirds of the total production in inland fisheries. In both of these regions, inland fish catches play a critical role in ensuring food security. Interestingly, India has recently surpassed China to become the leading producer of freshwater fish, marking the first time since the mid-1980s. India achieved an impressive production volume of 1.8 million metric tonnes in inland fisheries.

A record amount of 122.6 million tonnes of aquaculture products were produced worldwide in 2020, including 87.5 million tonnes of aquatic animals valued at USD 264.8 billion and 35.1 million tonnes of algae. While a significant portion of 68.1 million metric tonnes originated from marine and coastal aquaculture, approximately 54.4 million metric tonnes were cultivated within inland water systems. All areas saw continued growth in aquaculture in 2020, with the exception of Africa. Top producers Chile, China, and Norway were responsible for the expansion in their respective regions. However, the continent's two largest aquaculture producers, Egypt and Nigeria, had a fall, while the rest of Africa saw a growth rate of 14.5% compared to 2019.

Asia maintains its dominant position in the field of aquaculture, accounting for over 90% of the worldwide output. The share of aquaculture in global aquatic animal production reached an all-time high of 49.2% in 2020. Marine animal aquaculture has remained more successful than that of aquatic creatures not fed. Despite the enormous range of aquatic species raised for food, the output of aquaculture was dominated by a limited group of "staple" species, namely Global inland aquaculture will use grass carp, while marine aquaculture will use Atlantic salmon.

The Food and Agriculture Organization (FAO) plays a crucial role in overseeing the status of fishery resources. Their comprehensive, long-term examination of marine fishery stocks has yielded concerning findings. According to their research, there is a continuous decline in the abundance of marine fisheries resources. As of their latest assessment, only 57.3% of fishery stocks are being maximally sustainably fished, while a mere 7.2% fall into the category of under fished stocks. This marks a significant shift from the situation in 1974 when 90% of fisheries stocks were considered to be at biologically sustainable levels. By 2019, this proportion had dropped to 64.6%. Despite these alarming trends, there is a silver lining. In 2019, biologically viable stocks constituted 82.5% of the total aquatic product landings. This represents a 3.8% increase from 2017, indicating some progress in the face of deteriorating conditions. To illustrate this further, consider that, on average, 66.7% of the stocks of the ten most harvested species in 2019, which include species like Anchoveta, Alaska pollock, skipjack tuna, Atlantic herring, yellowfin, blue whiting, European pilchard, Pacific chub mackerel, Atlantic cod, and oversized head hair-tail, were fished within biologically sustainable levels. This percentage is slightly higher than the figure recorded in 2017. This demonstrates the more effective management of these extensive stocks. The potential benefits of this effective management are substantial. Rebuilding overfished stocks could result in a remarkable increase in the productivity of marine capture fisheries, estimated at 16.5 million tonnes. This, in turn, would have far-reaching positive effects on food security, nutrition, local economies, and the well-being of coastal populations.

Due to the constrained personnel and financial resources allocated to monitoring and management, critical inland fisheries have difficulties in many least developed and developing nations. This deficiency of resources is a major barrier. Inland fisheries have a lower profile, even in some affluent nations. Consequently, due to competing demands, stock assessment and monitoring may not receive the same level of attention. To tackle this issue in 2016, the Food and Agriculture Organization (FAO) initiated the development of a global threat map for inland fisheries. This map aims to establish a foundational measurement framework for tracking alterations in critical basins and enhancing the management of inland fisheries. Based on initial findings from this map, it is observed that 55% of inland fisheries across major basins are experiencing moderate stress, while 17% are facing significant pressure.

## Farmers Involved in Fisheries Sectors and the Status of Fishing Craft and Gears

Based on forecasts, it is anticipated that in the year 2020, the global count of fishing vessels will reach approximately 4.1 million, marking a decrease of 10% in comparison to the figures from 2015. This trend underscores the concerted efforts of several nations, most notably China and various European countries, to scale down the overall size of the worldwide fishing fleet. Notably, Asia continues to maintain its status as the region with the largest fishing fleet, boasting control over roughly two-thirds of the entire global fleet. Additionally, it's noteworthy that more than 75% of the 2.5 million motorized containers across the globe are concentrated in Asia, while a substantial 97% of non-motorized containers are primarily used for transportation between Asia and Africa.

According to estimates, 58.5 million people were employed full- or part-time in the primary production industry in fisheries and aquaculture in 2020. While the number of anglers worldwide has decreased in recent years, about 35% of people were working in aquaculture, 84% of all fishermen and fish farmers lived in Asia in 2020. In the primary industry as a whole, women made up 21% of the workforce (18% in fisheries and 28% in aquaculture). Even yet, they only make up 15% of full-time workers in aquaculture and fisheries in 2020, and their employment tends to be more erratic. However, even taking into consideration the data that is available for the processing industry, women made up 71% of part-time jobs and slightly over 50% of full-time jobs.

## Consumption Status of Fish

In recent times, there have been significant transformations in both the utilization and processing of fisheries and aquaculture products. Specifically, in 2020, approximately 89 percent, equivalent to 157 million tonnes, of the global output (excluding algae) found its way into direct human consumption, marking a substantial increase from the 67 percent recorded in the 1960s. The remaining portion, totaling more than 20 million tonnes, was allocated for non-food purposes, with a predominant share being directed towards fishmeal and fish oil production, while the rest was utilized for various purposes such as pet food, ornamental fish, bait, pharmaceutical applications, cattle and fur production, and direct aquaculture feeding.

Concerning direct human consumption, the majority of aquatic food products (excluding algae) were primarily consumed in live, fresh, or cold forms, closely followed by frozen, prepared, preserved, and cured variations. Notably, in Asia and Africa, a larger proportion of aquatic food production is preserved through methods such as salting, smoking, fermenting, or drying, compared to international standards. Additionally, there has been a growing trend in the utilization of by-products, which serve both food and non-food purposes. For instance, by-products played a significant role, contributing to over 27 percent of global fishmeal production and 48 percent of fish oil production.

The global consumption of aquatic food, excluding algae, experienced a consistent annual growth rate of 3.0 percent from 1961 to 2019, which is nearly twice the rate of annual population growth, standing at 1.6 percent. Consequently, per capita consumption reached a record high of 20.5 kilograms in 2019. However, it is worth noting that preliminary forecasts indicate a decline in consumption in 2020, primarily due to reduced demand triggered by the COVID-19 pandemic. Subsequently, a modest increase is anticipated in 2021. The average amount of fish consumed in India is 6.31 kilogramme per person year (Fisheries Statistics Handbook, 2022). the state-by-state fish consumption per person for India in 2020–2021. The figures represent the typical annual intake of fish per person for each state and union territory in kilogrammes (kg).

| **State-wise fish consumption per capita 2020-21** | |
|---|---|
| **State/UT** | **Per capita consumption(per capita per kg)** |
| Andaman and Nicobar Islands | 77.84 |
| Andhra Pradesh | 8.91 |
| Arunachal Pradesh | 3.65 |
| Assam | 11.89 |
| Bihar | 9.6 |
| Chandigarh | Not received |
| Chattisgarh | 19.7 |
| Dadra and Nagar Haveli and Daman and Diu | 1.2 |
| Delhi | 0.25 |
| Goa | 78 |
| Gujarat | 7.44 |
| Haryana | 0.3 |
| Himachal Pradesh | 2.22 |
| Jammu and Kashmir | 6.00 |
| Jharkhand | 10.61 |
| Karnataka | 10.55 |
| Kerala | 17.93 |
| Ladakh | 0.05 |
| Lakshadweep | 125 |
| Madhya Pradesh | 4.87 |
| Maharashtra | 4.72 |
| Manipur | 18.25 |
| Meghalaya | 9.00 |
| Mizoram | 3.38 |
| Nagaland | 6.06 |
| Odisha | 16.34 |
| Puducherry | 18.88 |
| Punjab | 0.4 |
| Rajasthan | 0.01 |
| Sikkim | 1.16 |
| Tamil Nadu | 9.3 |
| Telangana | 8.7 |
| Tripura | 25.53 |
| Uttarakhand | 0.81 |
| Uttarpradesh | 11.09 |
| Westbengal | Not received |
| All India | 6.31 |

*Source: Fisheries Statistics Handbook, 2022*

## Marine Fisheries in India

The extensive and diverse fisheries resources of India, which include deep seas, lakes, ponds, and rivers, are home to more than 10% of the world's fish and shellfish species. Marine fisheries resources may be found throughout the country's large coastline, in its sizeable Exclusive Economic Zone (EEZ), and on its expansive continental shelf.

The maritime resources of the nation include an EEZ of 2.02 million square kilometres, a continental shelf of 0.53 million square kilometres, and a coastline of 8,118 km. According to estimates, the marine fisheries potential in Indian seas is 5.31 MMT, consisting of around 43.3% demersal, 49.5% pelagic, and 4.3% oceanic groups, 1.1% island resources, and 1.8% other resources. Andaman & Nicobar Islands have a coastal length of 1912 km, with 51 landing centres. Andhra Pradesh has a coastal distance of 974 km and the highest number of landing centres, with 350. Dadra & Nagar Haveli and Daman & Diu have a coastal length of 27 km and eight landing centres.

**Table – State-wise coastal length and landing centres**

| State and Union territory | Coastal length | Landing centres |
|---|---|---|
| Andaman & Nicobar island | 1912 | 51 |
| Andhra Pradesh | 974 | 350 |
| Dadra & Nagar Haveli and Damman & Diu | 27 | 8 |
| Goa | 104 | 32 |
| Gujarat | 1600 | 107 |
| Karnataka | 300 | 115 |
| Kerala | 590 | 174 |
| Lakshadweep | 132 | 20 |
| Maharashtra | 720 | 173 |
| Odisha | 480 | 55 |
| Pondicherry | 45 | 22 |
| Tamil Nadu | 1076 | 301 |
| West Bengal | 158 | 49 |
| **Total** | **8118** | **1457** |

Gujarat has the longest coastline among the states, with 1600 km, accompanied by 107 landing centres. Tamil Nadu follows with a coastal length of 1076 km and 301 landing centres. Other states like Maharashtra, Kerala, Karnataka, and Odisha have substantial coastal sizes and many landing centres. The total coastal length of all the states and union territories combined is 8118 km, with 1457 landing centres.

# Total Marine Production State-wise

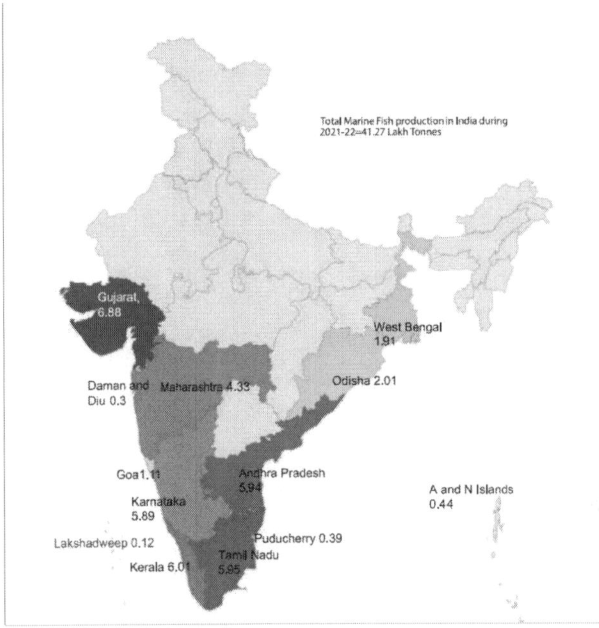


**Fig. 1. Marine fisheries production during 2021-2022**

*(Source: Handbook on Fisheries Statistics 2022)*

**Table -The Potential of fisheries resources in the Indian Exclusive Economic Zone**

*Source (Fishery Survey of India, 2018)*

| A. Conventional resources | Quantity in tonnes |
|---|---|
| Demersal resources (mainland) | 2298281 |
| Pelagic resource (mainland) | 2631827 |
| Lakshadweep (excluding oceanic) | 14490 |
| Andaman and Nicobar(excluding oceanic) | 43794 |
| Oceanic (for the entire EEZ) | 230832 |
| Others | 91369 |
| Subtotal | 5310593 |
| **B. Non-conventional resources(additional potential)** | |
| Deep sea Myctophids | 1000000 |
| Oceanic squids | 630000 |
| Jellyfish | 200000 |
| Marine microalgae | 17775 |
| Subtotal | 1847775 |
| Resource total grand | 71,58,368 |

## Inland Fisheries Resources and Fish Production

Rivers, canals, floodplain lakes, ponds, tanks, reservoirs, brackish water, saline/alkaline impacted regions, etc. are some of the resources used by inland fisheries. Inland Fisheries has expanded in absolute terms, but not at the rate it could otherwise increase. The vast interior resources consist of 0.27 million km of rivers and canals, 2.45 million ha of ponds and tanks, 1.2 million ha of floodplain lakes, and 3.15 million ha of reservoirs. In India, inland fish production would total 121.21 lake tonnes in 2021–2022.

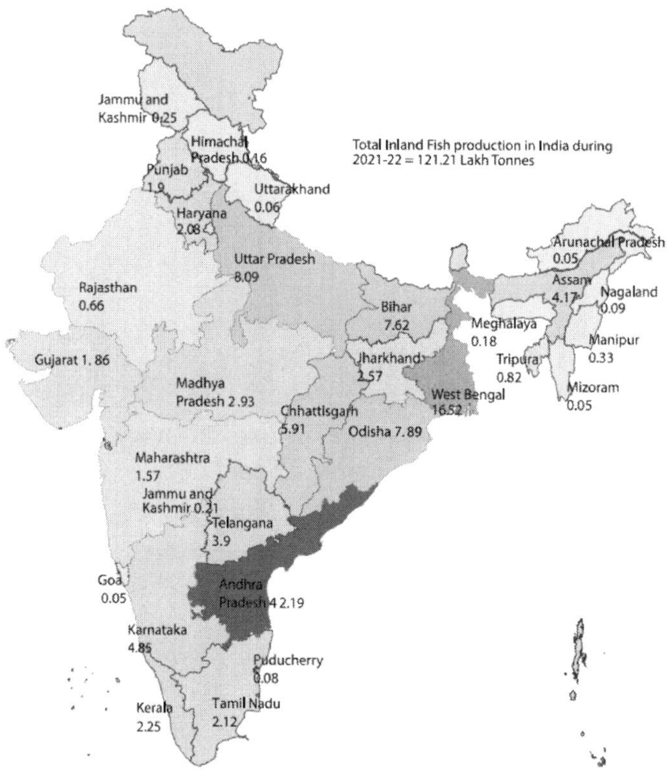


**Fig. Total inland fish production**

*(Source: Handbook on Fisheries Statistics 2022)*

## References

DoF, 2022. Handbook of Fisheries Statistics 2022. Department of Fisheries, Govt. of India.

SOFIA, 2022. The State of World Fisheries and Aquaculture 2022. FAO | Food and Agriculture Organization of the United Nations.

CHAPTER

# Cultivation System

# 2

The primary objective of fish culture is to generate a plentiful supply of high-quality fish intended for human consumption. The specific approach employed for fish culture in a region is contingent upon the accessible water source and prevailing climatic conditions. Consequently, fish are cultured using diverse methodologies across various parts worldwide. The most commonly used methods are 1. Cage culture, 2. Pen culture, 3. Recirculatory water system (Bandyopadhyay 2022).

## Cage Culture

- In many regions worldwide, the primary water source is flowing water or large bodies of water where diverting the water into a pond is not feasible. In such cases, fish can be cultivated in small cages within these waters. This method is commonly employed for carp cultivation in limited water areas. The cages can be rectangular boxes or bamboo cylinders that can float in the water current, allowing water to pass through.
- Rectangular cages can be constructed using materials such as bamboo, wire screens, or nylon nets with a mesh size of 10mm. These cages, made of bamboo or woody material, are enclosed by nylon nets on all sides. Select cage sites protected from heavy winds to prevent any damage to the cages and fish mortality due to imbalances (Abery *et al.*, 2005).
- Floating cages are utilized in Japan and Russia for carp culture within lakes, while Thailand has employed cages for fish cultivation for many years. Traditional methods of cage cultivation are practised in

Java and Cambodia for tilapia and *Clarias batrachus* (walking catfish) cultivation, respectively. In India, cage culture is adopted in Tripura and West Bengal.

- Fast-flowing water is ideal for cage culture, as slow-moving water can lead to issues like oxygen depletion and competition for food among fish. Continuous water flow in cages facilitates the removal of waste materials and maintains oxygen levels. Moreover, fish cultivation using this method requires a relatively small budget.
- Cages with dimensions of 2x1x1 meters are suitable for small-scale fish cultivation, whereas commercial culture may require larger cells measuring 5x4x2 meters. Cages are also advantageous for producing hybrid strains by introducing selected pairs of male and female fishes. Additionally, stocked fishes are protected from predators when confined within cages (Ginty and Rakocy 2007).

**Fig. Cage Culture**

*(Source: https://images.app.goo.gl/nFiRwmiMKAFabCHp6).*

## Advantages

Fish cultivation can be successfully carried out in both small water areas and flowing water environments. This form of aquaculture offers several advantages:

- **Feasibility in limited spaces:** Fish farming can be conducted in small water areas, making it accessible to individuals or communities with limited land resources.
- **Cost-effective:** Fish cultivation typically requires a smaller budget compared to other forms of animal farming. This afford ability makes it more attainable for aspiring fish farmers.

- **The natural environment in flowing water:** In the case of cultivating fish in flowing water, a natural atmosphere is maintained. This ensures that the fish have access to an environment similar to their natural habitats, promoting their overall health and growth.
- **Protection from predators:** When fish are stocked in controlled environments, such as ponds or fish tanks, they are shielded from potential predators. This protection increases the chances of survival and allows for better fish population control (Huang *et al.*, 2011).
- **Hybridization opportunities:** Fish cultivation also provides the chance for hybridization, which involves breeding different species of fish to create hybrids with desirable traits. This allows for producing new and improved fish varieties that may have enhanced growth rates, disease resistance, or other beneficial characteristics.

## Disadvantages

- The provision of food necessitates significant financial outlay.
- The cages are vulnerable to damage and destabilization caused by strong winds and floods.
- The mortality rate of fish is high due to cage damage and instability.
- Continuous monitoring and adept handling are essential.

# Pen Culture

- This fish culture method is commonly practised in Scotland and Israel.
- Nowadays, it is being adopted in some Asian countries.
- Fish are cultured in pens located inside lakes or offshore areas with flowing water.
- These pens are enclosures made of bamboo or wooden poles and nylon nets.
- The poles are fixed in the bottom of shallow water areas, and nets are strung from pole to pole to form the enclosure.
- The nets are anchored at the bottom to prevent stocked fish from escaping.
- Nylon netting is used for the pens to ensure it doesn't rot.
- The pens' size varies, typically 20 meters in length, 10 meters in breadth, and 5 meters in depth.

- If the water is not fertile, the stocked fish is provided supplementary feeding.
- In fertile waters, fishes grow faster, eliminating the need for additional food supply.
- Pens require minimal fertilization and maintenance.
- However, constructing pens can be expensive.
- The cost of building a pen is equivalent to constructing a pond of the same size.
- The lifespan of a fish pen in water is typically 3 to 5 years (Bandyopadhyay 2022).

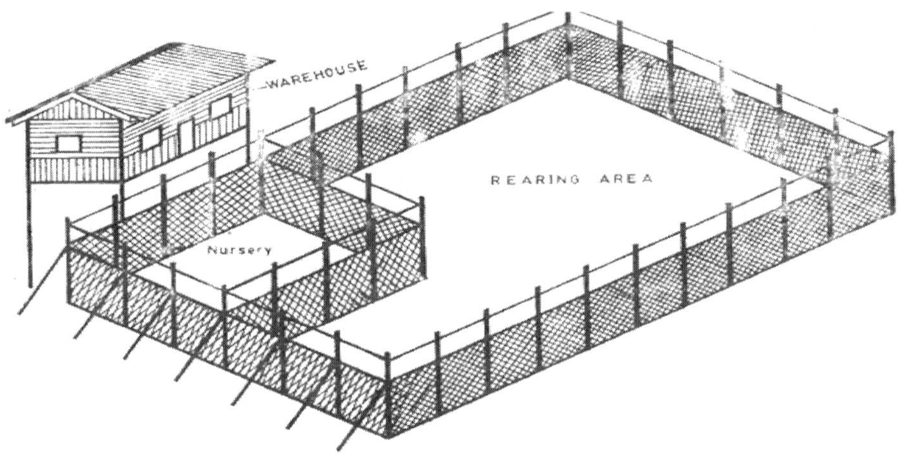


**Pen Culture**

*(Source: https://images.app.goo.gl/gNuinEqU83GUQrqV9)*

## Advantages

- Fish cultivation is feasible in flowing water, allowing for efficient utilization of natural water resources.
- No fertilizer or manure is needed for fish cultivation, reducing the reliance on artificial inputs.
- There is no requirement for additional feeding, as fish can thrive on natural food sources available in the water body.
- Fishes are shielded from predators or enemies, providing a safer environment for their growth.
- Fish cultivation necessitates minimal maintenance, making it a relatively hassle-free practice.

- Fishes exhibit rapid growth rates in this system, enabling higher yields within shorter time frames.
- Overall, fish cultivation in flowing water offers numerous advantages, such as reduced dependency on external inputs, improved protection for the fish, and higher growth rates (Nunoo *et al.*, 2014).

## Disadvantages

- The construction of pens can be costly.
- Fish pens have a shorter lifespan compared to ponds.
- In sterile water, additional expenses are needed for supplementary feeding.
- It diminishes natural productivity.

## Recirculating Water System

- This rearing method is designed for fish farming in little water and small spaces.
- It employs a water circulation and recirculation system to enable prolonged use of the same water.
- The concept of recirculating water systems was first successfully developed by I. Motocava in Japan in 1951.
- The necessary components for this method include one fish tank, two filtration tanks, connecting pipes, a water supply, and an aeration facility.
- Aeration is crucial to maintain the oxygen level in the water since it is reused for an extended period.
- To remove waste accumulated at the bottom of the fish tank, water is transferred through connecting pipes to the filtration tank and then back into the fish tank.
- Following this process, the same water can be used until its maximum utility is reached, after which it is replaced with fresh water.
- Two filtration tanks are utilized to ensure a continuous supply of clean water.
- The filtration tanks are used alternately in the system.
- The tanks undergo a thorough cleaning process utilizing compressed air directed through a pipe system.

- By employing this technique, the yield of fish achieved consistently surpasses any other method used in fish cultivation (Xiao *et al.*, 2019).

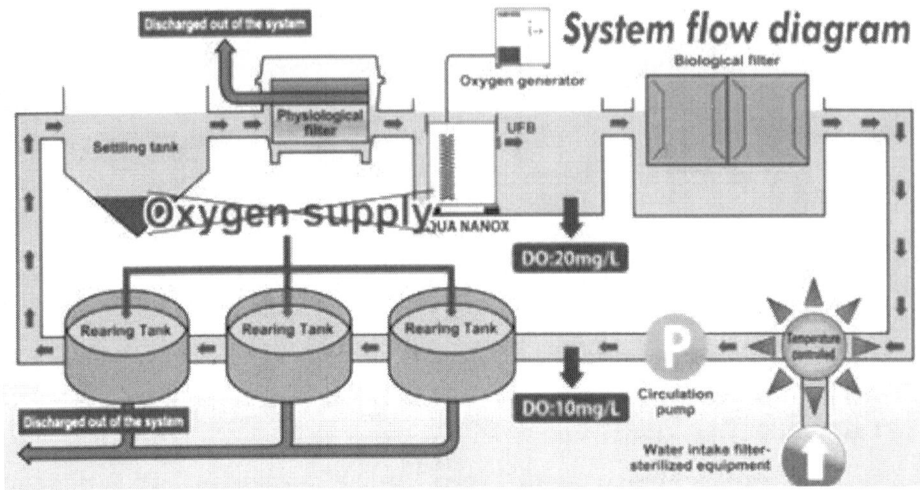


**Fig. RAS System**

*(Source: https://images.app.goo.gl/YnHvW4yxeFFrnu9j6)*

## Merits

- In this method, fish are cultured in a confined space called a fish tank or pond. The cultivation area is relatively small compared to traditional fish farming methods. The fish are raised in little water, providing a controlled environment for their growth and development.
- One of the advantages of this method is that water quality can be carefully maintained. By monitoring and regulating the water parameters such as temperature, pH levels, and oxygen content, the same water can be reused for an extended period. This reduces the need for frequent water changes and conserves water resources.
- Additionally, employing this method minimizes the expenditure associated with human labour. The confined space allows for efficient monitoring and management of the fish, reducing the need for extensive manual work. This results in cost savings for fish farmers.
- Another benefit is that the fish are protected from natural predators in open water bodies. By confining the fish in a controlled environment, such as a tank or pond, they are shielded from potential threats like larger predatory fish, birds, or other aquatic predators. This protection helps ensure the stocked fish's safety and well-being (Sharrer *et al.*, 2007).

- Furthermore, maintaining an aquatic environment in this method significantly reduces the mortality rate of stocked fish. By carefully controlling the water conditions, providing appropriate nutrition, and minimizing stress factors, the fish experience improved survival rates. This leads to higher overall productivity and profitability for fish farmers.
- Cultivating fish in a small space offers various advantages, including controlled water quality, cost-effective operations, predator protection, and reduced mortality rates. These factors contribute to the success and sustainability of fish farming using this method.

### Demerits

- The provision of artificial food demands substantial financial investment.
- Continuous aeration is a necessity.
- Implementation of this method is impractical in rural areas with a predominantly illiterate population.
- The presence of filtration facilities is essential.

## Earth-fill Raceway

- In earth-fill raceways, purely earthen ponds are utilized for various purposes, such as fish cultivation. These ponds are designed to be long and narrow and constructed using natural earth materials. The dimensions of these ponds typically measure 30.5 meters in length, 1.5 meters in depth, 9 inches wide at the surface, and 1.3 meters at the bottom.
- Multiple ponds of this type, usually between 4 and 8, are built to accommodate the desired fish population. One of the advantages of using this type of pond is that supplementary food is not required, as the ecosystem within the pond can sustain the fish naturally. These ponds possess a self-purification capacity, which helps in maintaining water quality. However, it is necessary to conduct regular maintenance to control vegetation growth along the pond banks.
- Due to the water content, the corners of the ponds can sometimes break, requiring annual repairs to maintain their structural integrity. Although these ponds can self-purify, the bottom may remain unclean, and toxic gases can be released. Therefore, it is essential to periodically clean the pond bottoms to ensure a healthy environment for the fish.

- One common application of these ponds is the cultivation of trout fish. These fish thrive in such environments and can be successfully raised within these earth fill raceway ponds (Sompech *et al.*, 2012).

## Advantages

- It possesses the ability to cleanse itself.
- The second statement can be rephrased as: "Since it relies on the natural growth of food within the pond, there is no need for any artificial food supply.
- The third statement can be restated as: "Aeration is unnecessary."
- The fourth statement can be paraphrased as: "Frequent cleaning is unnecessary."

## Disadvantages

- To prevent plant growth, utmost maintenance is necessary.
- Ponds necessitate yearly repairs.
- Vegetation thriving along the pond banks can pose a threat to stocked fish.
- The unclean bottom of the pond may release toxic gases that negatively impact fish growth.

## Concrete Raceway

- In this type of fish cultivation, cemented (concrete) tanks are used. The concrete tanks have the following dimensions: 30.5 meters in length, 1.1 meters in depth, 6 meters in breadth at the top, and 3 meters at the bottom. Typically, a series of 6 to 10 ponds are constructed.
- In concrete tanks, aeration is necessary, and it can be achieved by using a water pump. The flow of water required for a series of 6 ponds is 84 litres per second, while for a series of 10 ponds, it is 40 litres per second (Huggins *et al.*, 2005).
- To facilitate the feeding of fishes, a road should be prepared between the ponds. Additionally, keyways are provided in the concrete walls separating the ponds. These keyways contain waste collecting boards called check boards.
- The check boards play a crucial role in preventing the waste material from being carried over to another pond due to water flow. Instead, the waste material accumulates in the check boards. It can

then be collected and removed separately from the checkboard (Huggins *et al.*, 2005).

## Advantages

- Yearly repairs are unnecessary as there is no need for frequent maintenance.
- Since plants do not grow along the pond banks, fish are not exposed to any harm.
- These ponds are easily manageable in terms of disease control and disinfection.

## Disadvantages

- Fishes should be supplied with artificial food as natural food is absent.
- Frequent cleaning of the pond is necessary.
- Aeration is required to maintain the dissolved oxygen content of water.
- It does not have a self-purification capacity.

## References

Abery, N. W., Sukadi, F., Budhiman, A. A., Kartamihardja, E. S., Koeshendrajana, S., Buddhiman and De Silva, S. S. (2005). Fisheries and cage culture of three reservoirs in west Java, Indonesia; a case study of ambitious development and resulting interactions. *Fisheries Management and Ecology*, 12(5), 315-330.

Bandyopadhyay, B. K. (2022). *Freshwater Aquaculture: A Functional Approach.* CRC Press.

Huang, Y. C., Hsieh, H. J., Huang, S. C., Meng, P. J., Chen, Y. S., Keshavmurthy, S. and Chen, C. A. (2011). Nutrient enrichment caused by marine cage culture and its influence on subtropical coral communities in turbid waters. *Marine Ecology Progress Series*, 423, 83-93.

Huggins, D. L., Piedrahita, R. H. and Rumsey, T. (2004). Analysis of sediment transport modelling using computational fluid dynamics (CFD) for aquaculture raceways. *Aquacultural Engineering*, 31(3-4), 277-293.

Huggins, D. L., Piedrahita, R. H. and Rumsey, T. (2005). Use computational fluid dynamics (CFD) for aquaculture raceway design to increase settling effectiveness. *Aquacultural Engineering*, 33(3), 167-180.

McGinty, A. S. and Rakocy, J. E. (2007). *Cage culture of tilapia.* Oklahoma Cooperative Extension Service.

Nunoo, F. K. E., Asamoah, E. K. and Osei-Asare, Y. B. (2014). Economics of aquaculture production: A case study of pond and pen culture in southern Ghana. *Aquaculture Research*, 45(4), 675-688.

Sharrer, M. J., Tal, Y., Ferrier, D., Hankins, J. A. and Summerfelt, S. T. (2007). Membrane biological reactor treatment of a saline backwash flow from a recirculating aquaculture system. *Aquacultural Engineering*, 36(2), 159-176.

Sompech, K., Chisti, Y. and Srinophakun, T. (2012). Design of raceway ponds for producing microalgae. *Biofuels*, 3(4), 387-397.

Xiao, R., Wei, Y., An, D., Li, D., Ta, X., Wu, Y. and Ren, Q. (2019). A review of the research status and development trend of equipment in water treatment processes of recirculating aquaculture systems. *Reviews in Aquaculture*, 11(3), 863-895.

CHAPTER

# 3

# Site Selection of Fish Farm

## Introduction

In the realm of fish, many individuals perish in the natural environment due to ailments, insufficient oxygen, or becoming prey to other organisms or predators before reaching maturity. However, pond fish culture endeavours to counteract this scenario, aiming to generate a surplus of fish. By effectively managing predators within ponds, fish production can surpass that of their counterparts in natural bodies of water. Fish growth in ponds primarily occurs because they are confined and subjected to planned nourishment, reproduction, development, and harvest processes.

Fish farming primarily takes place in man-made structures that mimic small, shallow water bodies, such as ponds. These artificial ponds exhibit distinct characteristics that set them apart from natural lakes. They typically boast a relatively expansive littoral zone, which is the shallow area near the shore, and a restricted profundal zone, which is the deeper, central region of the water body. Additionally, the source of water for these ponds can vary (Francová *et al.*, 2019).

## Site Selection

Selecting the appropriate location for fish ponds is paramount when planning fish farms. The significance of the pond's positioning cannot be overstated, as it holds the potential to surpass the productivity of the surrounding land. Various factors necessitate consideration during the selection process, encompassing the desired quantity and classification of ponds, the topographical attributes of the designated area, the availability of a reliable water source, and the specific species of fish intended for cultivation. Following Pillav's research (2008), a comprehensive evaluation of these factors is imperative for making an informed decision regarding the optimal location for the fish ponds.

Poor agricultural land has the potential to be transformed into highly productive fish farms. The critical factor for achieving large quantities of fish is the presence of healthy soil. In cases where a pond is constructed on agricultural land that fails to yield good crops, proper maintenance can lead to a significant improvement. Over time, the soil at the bottom of the pond will become more fertile than its original state.

After the fish harvest, the pond can be replanted with a land crop, such as corn, and allowed to grow. Once the corn is harvested, the area can be converted into a fish pond. This approach, as described by Jadrnicek and Jadrnicek (2016), enables the utilization of the land for two economically viable crops—fish and maize. Furthermore, fish can be cultivated with paddy in paddy fields, allowing for dual-purpose land use. The decision regarding the land's use becomes crucial in such cases. It is worth noting that ponds can sustain fish rearing throughout the year, unlike other options.

## Criteria for site selection:

- The presence of land in a continuous, adequately shaped, ideal-sized plot with all facilities.
- Either surface water or groundwater of sufficient quality should be available at the site.
- The location's water and soil must be appropriate for fish culture.
- Floods shouldn't occur at the location.
- The site should have reasonable access roads and transportation options.
- The location must have the phone and electrical connections.
- The fish seed ought to be quickly and abundantly available there.
- Marketing facilities should be accessible close to the site.
- The location ought to be remote from populated areas.
- A drainage system ought to be connected to the location.
- The location ought to be far from contaminated areas.

The significant elements that collaboratively contribute to creating an ideal location for a fish pond include the following:

## Water Supply

The primary determinant in choosing a site of utmost importance is the presence of water. Fish rely solely on water for their nutritional needs, making its accessibility pivotal. An ideal site swiftly satisfies its initial criterion if it consistently possesses a year-round water supply. Even in cases where the natural water source experiences periodic shortages, a site might remain under consideration if there exists a feasible means of water storage. However, it's worth noting that water availability cannot always be assumed as a constant factor.

The requirement that water be always accessible and plentiful is the most crucial one. Near the site, there needs to be a dependable water supply. Having an ample and consistent water source is crucial for the proper maintenance of pond water levels, especially in the context of carp culture. Commonly utilized water sources for carp culture ponds encompass rivers, streams, springs, canals, and the surface runoff resulting from precipitation. The suitability of water from any of these sources for fish culture depends on its freedom from contamination. It is important to note that various natural water sources are available in India (Cosgrove *et al.*, 2015).

- **Water from natural sources:** Many ponds are typically supplied with water sourced from natural springs or redirected from nearby rivers, streams, or lakes.
- **Springs:** Some ponds are constructed near springs that provide the water for the ponds. Since spring water is uncontaminated and free of unwanted fish and fish eggs, it is an excellent source for fish culture. Spring water is obtained from underground sources. Filtering spring water might become a crucial step when it has traversed a considerable distance, prior to its suitability for application in a fish pond.
- **Rainfall:** Some ponds, called "sky ponds," get all their water from rainfall.
- **Runoff:** In certain instances, ponds may take the form of sand and gravel pits which become inundated with water as a result of the flow of precipitation and surface water from the surrounding land area.
- **Wells:** Utilizing well water presents the optimal choice for supplying water to a fish pond (Chakroff, 2018).

Wells supply water, but they often lack oxygen, which fish need. The oxygen levels in ponds can increased by stirring, beating with bamboo sticks, or using small motors (Somerville *et al.*, 2014).

The majority of the time, the fish pond is filled with water from rivers, streams, or lakes. A diversion channel has been built to connect these two points, facilitating the transfer of water from its source to the ponds. This approach proves to be an efficient means of pond filling due to its straight forward water management. A gate or plug can block the channel when the pond is full, stopping the water flow into the pond (Bowman *et al.*, 2007).

Problems with this water supply:

- Excessive rainfall in tropical regions during the wet season can pose a significant risk to ponds, and it is imperative to manage the surplus water effectively.
- Flooded pond leads to fish escaping and the pond becoming empty.
- Unfiltered water introduces unwanted fish and their eggs into the pond.
- Clearwater from the source may lack sufficient nutrients, requiring fertilizers.
- Muddy water must settle before entering the pond, requiring a separate settling area.
- Bright green water indicates a high concentration of fish food organisms.
- Dark water may contain acid and requires the addition of lime.

## Topography

Describe the terrain's shape, including level, hilly, upland, lowland, etc. The kinds of ponds that can be built depend on the land's topography. The land's topography and the farmer's needs determine the pond's location, shape, and size. The topography that enables water to fill and drain ponds using gravity is the most advantageous for fish ponds. Ponds that are situated on an incline can be rapidly drained. In contrast, when ponds are located on level ground, it becomes necessary to incorporate an internal slope within their structure. This internal slope facilitates the natural drainage process driven by gravity. However, if the ponds are positioned on flat land without this slope, a pump becomes essential for efficient drainage (Burman *et al.*, 2013).

In general, the ponds should be flat or gently sloped towards the outlets. Each pond's construction costs, water intake, and outlet are based on its topography. The site must be selected so that, to the greatest extent possible, the earth made available through excavation balances the earth required to raise or fill dykes. Before designing and building, the site needs to be thoroughly surveyed to determine the topography and land configuration. (New and Semusk, 1982).

## Soil

The soil in the area is another crucial factor in site selection. The pond's soil needs to be capable of holding water. Due to its nutrients, it also helps the water be more fertile.

One with a lot of clay is the ideal soil for a pond. Water is well-retained by clay soils. If the soil exhibits a slippery and smooth texture, it is likely to possess a significant clay content. Conversely, if it feels gritty or rough to the touch, it probably contains a substantial amount of sand. Smooth-textured soil would be advantageous for the construction of a fish pond since it can retain water effectively, thanks to its higher clay content. In summary, the presence of clay makes the soil ideal for pond construction as it enhances water retention capacity (Tarolli *et al.*, 2016).

One reliable technique for assessing the suitability of soil for the construction of a fish pond entails the process of dampening a handful of soil with a modest quantity of water until it reaches a moist consistency. Following this, you can assess its suitability by applying pressure, as in gently squeezing it. If the soil maintains its form and structure after you release your hand, it is regarded as conducive for the establishment of a pond.

Ponds can also be built in sandy soils, but it takes more work, and sometimes it doesn't work. Only clay soils can be used to construct large ponds. Only small ponds are possible in rocky or shifting sand soil. The fertility of the pond depends on the topography as well.

Fertility pertains to the quantity of nourishment accessible within the pond for the fish and serves as an indicator of the nutrient levels present in the aquatic environment. Usually, there are plenty of fish food organisms in fertile ponds. Fe, Ca, and Mg are essential nutrients in the pond's soil. The acid found in soil is another component that is bad for fish. Following a heavy downpour, new ponds have occasionally seen high fish mortality. The presence of significant quantities of acids from the soil being transported into the ponds can be attributed to the substantial rainfall (Boyd, 1995). Whether the ground has been used to grow crops is an excellent way to determine its quality. The land will be suitable for the fish ponds if crops flourish there. Avoid using porous and peaty soil because it will neither hold water nor allow for compaction. The soil will experience significant percolation if it possesses organic qualities and is characterized by porosity. To assess the subsoil composition, it is essential to collect random samples from the designated area. This sampling will help ascertain the presence of a clay layer, approximately 1 to 1.5 meters thick, beneath the pond's bottom.

## The Quantity, Configuration, and Dimensions of Aquatic Reservoirs

The potential site determines how many ponds will be present. Ponds for nursery, rearing, and stocking should be on the site. The dimensions of ponds are influenced by the geographical features, the presence of water resources, and the requirements for their use. Due to the minuscule size of the fry, nursery ponds tend to be more compact in comparison to the larger rearing ponds. The most extensive ponds in the fish farm are stocking ponds, typically larger than rearing ponds (Janauer *et al.*, 2016).

Smaller ponds offer easy and quick harvesting, rapid draining and refilling, simple disease treatment, and resistance to wind erosion. Expansive water bodies of a greater size exhibit reduced expenses in their establishment, demand a smaller land footprint per hectare of aquatic expanse, harbor increased oxygen levels, and offer the flexibility to be alternated with rice or other agricultural cultivations. In the fish farm, smaller ponds are preferable to a few larger ones because larger ponds are more challenging to manage. The ponds should not be wider than 40 metres so that only a few fishermen can effectively harvest the fish. The operations will be simpler if the ponds are rectangular.

The fish that are being raised determines the pond's depth. Various fish species exhibit preferences for specific types of food, and the depth of ponds plays a pivotal role in determining the kinds of food sources available within them. To illustrate, consider the common carp as an example. This particular species thrives in ponds with a maximum depth of 2 meters, as it primarily sustains itself by consuming worms and various organisms that inhabit the pond's bottom. However, during the initial fry stage of their development, carp shift their dietary habits and exclusively feed on plankton as well as minuscule floating plants and animals suspended within the water column. Consequently, carp fry nursery ponds are typically quite shallow, with depths of approximately 0.5 meters, to ensure an ample supply of these tiny, floating organisms. Because sunlight cannot reach more profound areas of ponds, they will not produce as much food.

A shallow pond runs the risk of turning cloudy, being susceptible to takeover by aquatic vegetation, and heating up rapidly. Ponds and embankments can have inclines varying from a $1\frac{1}{2}$ horizontal to 1 vertical ratio to a 2 horizontal to 1 vertical ratio. The bottom of the pond should be slanted toward the outlet to facilitate proper drainage. Ponds should be equipped with regulated inflows and outflows to enable easy filling and draining. Deeper ponds are ideally located at lower elevations, minimizing the need for extensive earthwork (Grafton and Hussey 2011).

## Survey

The initial phase involves demarcating the intended location for the forthcoming fish pond construction, marking a pivotal starting point. A critical aspect of this endeavor lies in ascertaining the innate incline of the terrain where the primary enclosure will be erected. It is imperative to identify the steepest segment of the hillside, designating it as the foundation for the lower boundary of the pond's perimeter. This specific region will house the infrastructure for the pond's drainage system. Even in cases where the topography appears relatively level, there exists a subtle undulation that needs recognition. Therefore, the land undergoes a meticulous survey to pinpoint the direction of the slope before embarking on the construction process.

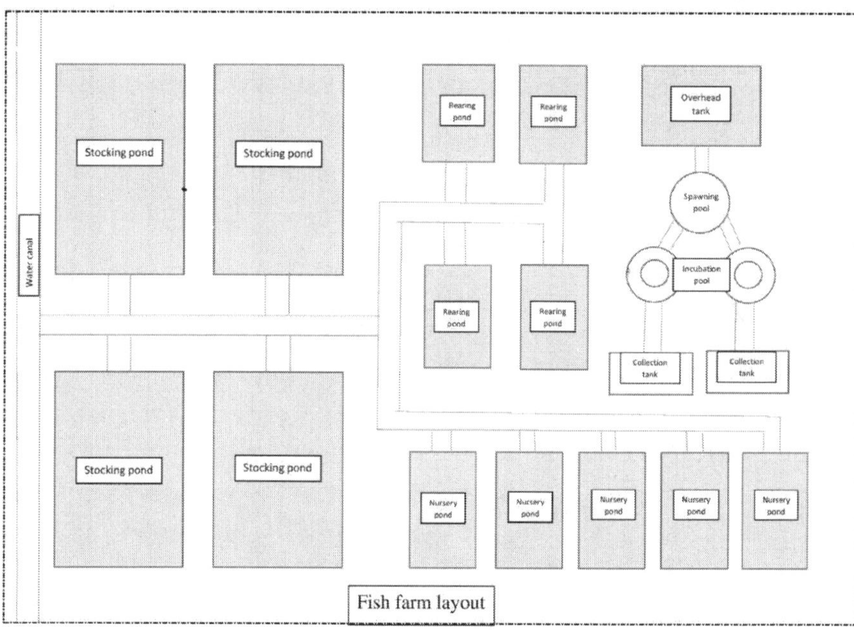


**Farm Designs and Layout**

*(Source: Authors)*

## Designing

Investigating survey findings, atlases, soil types, topography, water resources, etc. should be the first step in the design of fish ponds. The entire design and layout of ponds and dykes will be put into practise in accordance with the survey results. Planning where and how many nurseries, rearing ponds, and stocking ponds will be built is crucial when designing the fish farm.

A fish farm can only have nursery and rearing ponds built for seed production. The stocking pond has a small area set aside for breeding stock. Additional stocking ponds are built in a fish farm to grow fingerlings into table-sized fish. All three pond types (nursery, rearing, and stocking) are required in a mixed fish farm; the quantity of ponds should be based on the desired stocking density (Grafton and Hussey 2011).

## Construction

After the design, prepare itemized work estimates and approximate construction costs.

- Construction time
- Preparation the site
- Mark out the ponds
- Excavation of the pond
- Drainage system
- Sluice
- Water inlet
- Sealing the pond bottom

### 1. Construction time

The timing of pond construction is of utmost importance in effective pond management. Opting to complete pond construction during the summer season allows for the prompt utilization of the pond for agricultural purposes.

### 2. Preparation the site

Before commencing the construction process, it is essential to undertake site clearance procedures. This entails cutting and removing all bushes, small plants, and similar vegetation, including their roots. It is imperative to ensure that the roots are entirely extracted, as failure to do so may result in a potential leakage problem arising in the future. If there are any trees in proximity to the construction site, it is recommended to prune the branches in a manner that directs them away from the site. This will facilitate sunlight penetration into the area and prevent leaves from falling into the water, minimizing potential issues. While it is preferable to have trees near ponds, it is recommended that they be situated at least 5 meters away from the water's edge.

## 3. Mark out the ponds

Before cleaning the pond area, the outlines of ponds and dykes must be marked. Mark the main wall or dyke and any additional walls with stakes. The walls should be substantial. Plan the walls' height and depth for the pond. The walls must always be 50 cm and at least 30 cm higher than the water level for a small pond, respectively (Reid,2014).

## 4. Drainage System

In order to remove the water from the pond, a drainage system is put into operation. This system is composed of a network of drainage ditches that carry the water away from the pond, along with an outlet system designed to discharge the pond's water. Opting to construct the pond in an area with a favorable slope represents the most effective and uncomplicated method for establishing a reliable drainage system. It is important to note that certain components of the drainage equipment traverse through the walls; therefore, it is imperative to construct the drainage system prior to building the pond embankment (Muiret *al.*, 2010).

Inserting a bamboo or plastic tube into the base of the wall and directing it toward the center of the pond offers a straightforward means of emptying it. To prevent fish from accessing the interior of the tube within the pond, it's advisable to install a screen at that end. The opposite end of the tube should be sealed with either clay or wood. When it's time for harvest, the plug can be removed to facilitate pond drainage. Syphoning and using a pump represent alternative methods for emptying the ponds, although they are not as commonly employed. In a syphoning system, a rubber or plastic tube is secured, with one end submerged in the pond and the other outside, with the submerged end positioned lower than the inlet. This arrangement creates a vacuum in the tube, aiding in the water's removal from the pond. For those willing to invest, engine-powered pumps can be utilized to expedite the draining process, although they come with a substantial cost.

## 5. Excavation of the pond

Bulldozers or manual labour can both be used to complete the excavation. If a bulldozer is used, manual labour should provide the final shaping. To make a good drainage slope, pond sides and bottoms should be finished properly and trimmed. The pond bottom should typically slope by 2 to 5 percent. If the land for the pond is properly chosen, taking into account the natural topography, only a small portion of the pond bottom will need to be excavated. The most important element of a pond is that the bottom must slope to be drained. The main wall or dyke should be constructed on the lower side of the slope if the pond site has one. The excavated soil can be topped off and grass planted after the pond walls have been constructed. In this fertile topsoil, the grass will flourish and keep the walls from eroding. The pond's bottom must be cleared by removing minute rocks, roots, and other obstructions and stumps to stop the harvesting process from causing the nets to catch on and tear. The grass at the bottom of the pond won't need to be removed because it will rot and die once the water is added, adding nutrients to the water.

Since the top two feet of soil contain a lot of roots and other organic matter, it should be removed. After the topsoil has been removed, the core trench is cut. If the soil is porous, the seepage issue might develop later. The provision of a clay core would be necessary to stop seepage. The ideal ground is a blend of clay and sand. Pure clay soil is prone to leaks and cracks. Pure clay cannot be used alone; it must be combined with other soils. Humus, peaty soils, and turf should not be used (Mishra, 1994).

A pond's dykes need to be sturdy enough to withstand weather changes. Dike erosion in large ponds is a problem that necessitates routine attention. To stop the erosion of dykes, pitching made of brick and stone may be offered (Mishra, 1994). Earthen dykes can be protected from erosion with bamboo piling and bamboo jab free on top. Holes are another common problem that need to be immediately filled with stiff clay, lime, and cement, and then properly compacted. The earthen dykes will be more permanently guarded against crab or rat holes using concrete blocks, stones, or bricks.

The type of material used to build the embankment will determine how steeply its side slopes will be. Less slope should be present than angle. When the soil contains a substantial amount of clay, the outer surface of the wall can exhibit a steeper inclination compared to the inner side. Typically, when building an embankment, the exterior slope is designed with a ratio of 1:1, meaning that for every 1-meter increase in height, there is an accompanying 1-meter increase in horizontal distance. Conversely, the interior slope is set at a ratio of 1:2, indicating that for every 2 meters in height, there is a 1-meter change in horizontal distance. In simpler terms, a slope ratio of 1:2 signifies that the embankment's length increases by 1 meter for every 2 meters in height gained. Grass should be planted on the embankment once it has been built. The grassroots support the wall's structure and stop soil erosion. Trees shouldn't be planted on walls because as their roots spread, they will cause damage.

## 6. Sluicea

A screened gate positioned within a waterway leading to the sluice drainage portal permits the controlled release of water from the reservoir. The principal structure comprising the sluice gate for drainage within a pond is characterized by an upright sluice mechanism centrally located within the wall, its sides extending into the wall structure. This channel can be fashioned from various materials, including brick, cement, or wood, and may encompass one or two wooden gates that are adjustable to either admit or release water from the pond. Furthermore, to bar undesirable aquatic species from infiltrating the inlet and to prevent resident fish from exiting through the outlet, sluices incorporate a screened gate. Conversely, the monk, a distinct component, serves a similar purpose but differs from the sluice in that it remains detached from the pond's wall. While a sluice can be employed at the inlet, a monk is never deployed in this location. The monk's drainage system is instrumental in regulating the water level and ensuring that fish remain within the confines of the pond, thus averting their escape. (Chakroff, 1984).

## 7. Water inlet

Every pond needs a source of water input, unless they are naturally replenished by a spring or rainfall. Filters should be placed in the channel during inlet construction to prevent unwanted fish or other materials from entering the pond and to maintain clean water. You can establish uncomplicated water entry points by installing a bamboo conduit of an appropriate size through the barrier and extending it into the reservoir. It's crucial to ensure that the inlet tube is situated at an elevation above the water level.

A good filter is a wire screen that can be horizontal or vertical. The horizontal screen is great for letting water into the pond, with a part of it sticking out at the inlet (Muir and Lombardi, 2010). Vertical screens are an option too. A nylon mesh bag attached to the inlet pipe serves as a good filter. Another option is a sand and gravel filter, which is efficient and cost-effective but requires a small tank at the water inlet. Saran fiber filters are like horizontal wire screens below the water inlet but are pricey and need frequent cleaning.

The choice of the filter is determined following an assessment of the water source, as discussed by Muir *et al.* 2010. If the water is characterized by turbidity or an abundance of grass or leaves, opting for a wire screen filter is advisable. Conversely, if the water source lacks organic materials, a mesh bag filter is a suitable choice. In cases where the water contains unwanted fish and an increased amount of organic matter, the Saram, sand, and gravel filters prove to be the most effective.

Maintenance of these filters is essential. They should be periodically removed and cleaned using a brush and fresh water, or they can undergo a flushing process while maintaining a continuous water flow, commonly referred to as backwashing. Whenever water is introduced into the pond, it is crucial to ensure these filters are cleaned accordingly.

The pond bottom must be sealed as the final step in pond construction to prevent leaks. There is no need for special sealing if the soil contains more clay. Sand-filled bottoms should be sealed to keep the water in. A clay core lining over the pond's base is constructed to fill the bottom.

### 8. Sealing the pond bottom

Cement blocks are another option for sealing the pond bottom, but they are pricy. Another way to seal is to use a polyethene, plastic, or rubber sheet liner. Gley, also known as biological plastic, is yet another method created in the USSR Technique: Pond Bottom Preparation.

- Clean the pond bottom.
- Spread animal manure on the cleaned pond bottom.
- Place banana leaves, cut grass, or any plant material over the animal manure layer.
- Place a covering of soil over the organic material.
- Compact the layers thoroughly.
- Allow 2-3 weeks for the materials to decompose and settle.
- Fill the pond after the elapsed time.

## References

Boyd, C. E. (1995). Bottom soils, sediment, and pond aquaculture. *Springer Science and Business Media.*

Burman, D., Bandyopadhyay, B. K., Mandal, S., Mandal, U. K., Mahanta, K. K., Sarangi, S. K., and Sharma, D. K. (2013). Land shaping—a unique technology for improving the productivity of coastal land. Bulletin No. CSSRI/Canning Town/Bulletin/2013/02. *Central Soil Salinity Research Institute, Regional Research Station, Canning Town, West Bengal, India,*

Chakroff, M. (1984). Freshwater fish pond culture and management (No. 36). The Corps.

Cosgrove, W. J., and Loucks, D. P. (2015). Water management: Current and future challenges and research directions. *Water Resources Research*, 51(6), 4823-4839.

Francová, K., Šumberová, K., Janauer, G. A., and Adámek, Z. (2019). Effects of fish farming on macrophytes in peaceful carp ponds. *Aquaculture International,* 27, 413-436.

Grafton, R. Q. and Hussey, K. (Eds.). (2011). *Water resources planning and management.* Cambridge University Press.

Jadrnicek, S., and Jadrnicek, S. (2016). *The bio-integrated farm: a revolutionary permaculture-based system using greenhouses, ponds, compost piles, aquaponics, chickens, and more. Chelsea Green Publishing.*

Mishra, P. K. (1994). Integrated watershed management–an overview. *Pk Mishra,* 1.

Muir, J. F. and Lombardi, J. V. (2010). Grow-out systems—site selection and pond construction. In *Freshwater Prawns. Biology and Farming* (pp. 127-153). Blackwell Science Oxford, UK.

New, M. B., and Semusk, S. (1982). *Freshwater prawn farming* (No. 225). Daya Books.

Ngugi, C. C., Bowman, J. R., & Omolo, B. O. (2007). A New Guide to Fish Farming in Kenya.

Pillay, T. V. R. (2008). *Aquaculture and the Environment.* John Wiley and Sons.

Prosdocimi, M., Cerdà, A., & Tarolli, P. (2016). Soil water erosion on Mediterranean vineyards: A review. *Catena,* 141, 1-21.

Reid, C. (2014). *Falling Into Place: An Intimate Geography of Home.* Beacon Press.

Sharma, D. K. (2013). Land shaping—a unique technology for improving the productivity of coastal land. Bulletin No. CSSRI/Canning Town/ Bulletin/2013/02. *Central Soil Salinity Research Institute, Regional Research Station, Canning Town, West Bengal, India, 38.*

Somerville, C., Cohen, M., Pantanella, E., Stankus, A. and Lovatelli, A. (2014). Small-scale aquaponic food production: integrated fish and plant farming. *FAO Fisheries and aquaculture technical paper,* (589), I.

CHAPTER

# Pond Management

4

## Introduction

In India's aquaculture industry, carps hold a prominent position as the primary fish species, constituting a substantial 85% of the total production. This significant share encompasses both native carps, such as Catla, Rohu, and Mrigal, as well as exotic carps, including Silver Carp, Grass Carp, and Common Carp. The practice of carp farming has experienced a remarkable surge in popularity across the nation, driven by technological advancements that have propelled production levels from a modest 600 kg per hectare to an impressive 2,000 kg per hectare over the course of the past three decades. States like Andhra Pradesh, West Bengal, Punjab, and Haryana achieve higher yields of 6-8 tonnes/ha/year. Thriving carp culture requires proper breeding to produce healthy fingerlings and fry. Availability of desired species' seed at the right time is crucial. Nursery rearing cares for 72-96-hour-old spawns until they are 25-30 mm fry. These fries are then reared for 2-3 months in another pond to grow fingerlings of about 100 mm (Hussan *et al.*, 2018).

## Nursery Rearing of Fry and Fingerling

### Nursery pond management

In the context of rearing fry in nursery settings, it is preferable to utilize small ponds ranging from 0.02 to 0.10 hectares in size, and these ponds should have a depth of approximately 1.0 to 1.5 meters. There are primarily two main systems employed for this purpose: drainable or non-drainable earthen ponds and cement cisterns. The process of nursery rearing involves several steps, as described by Basavaraja in 2007.

**Figure 1: A) Wall and bottom schematic diagram of Three in one nursery fish pond structure B) Nursery pond , with water (left) and drained (right) (Source: Wang *et al.*, 2018).**

## Pre-stocking pond preparation

Elimination of aquatic vegetation: The first step in pond construction is the removal of aquatic weeds. In fish ponds, excessive vegetation growth is undesirable because it can harbour predatory and weedy fish and insects, reduce productivity, and make the netting and free movement of fish difficult. Because of their limited depth and diminutive dimensions, manual techniques are typically exclusively employed in nursery settings and rearing ponds. Aquatic weeds can be eliminated in larger ponds using mechanical, chemical, and biological methods.

## Eradication of predatory organisms

The techniques employed to eliminate predatory weed fishes and other organisms encompass draining and exposing the ponds to sunlight for drying, or the application of appropriate piscicides. Mahua oil cake application at a rate of 2,500 kg/ha-m is advised before stocking fish seed. In addition to being a piscicide, the oil cake decomposes into organic manure and increases natural productivity. Fish can be effectively killed by adding 350 kg/ha-m of commercial bleaching powder (30% chlorine) to the water. By incorporating 100 kilograms per hectare-meter of urea, administered 18 to 24 hours prior to the application of bleaching powder, it is possible to reduce the amount of bleaching powder required by half (Jayasankar *et al.*, 2018).

## Control of aquatic insects

Aquatic insects and their immature forms pose a significant threat to hatchlings in aquatic nurseries, as they not only prey upon the young fish but also vie for the limited available food resources. To effectively manage and reduce the population of these air-breathing insects in aquatic environments, a simple and cost-efficient method involves the application of a soap-oil emulsion. This emulsion can be created using inexpensive vegetable oil at a rate of 56 kilograms per hectare, mixed with one-third of its weight in any affordable soap. By employing this soap-oil emulsion, it becomes possible to control and diminish the presence of aquatic insects and their larvae, safeguarding the hatchlings and ensuring that they have access to the necessary food resources for their growth and development.

The emulsion can be made using kerosene at 100-200 litres, diesel at 75 litres, liquid soap at 560 ml, or detergent powder at 2–3 kg per hectare of water.

The favored naturally occurring sustenance for fish comprises plankton, which is cultivated by introducing fertilizers into culture ponds. Initially, the ponds designated for seed production undergo liming treatment after the elimination of undesirable predatory and weed fish species, contingent on the soil's pH levels. Subsequently, these ponds are subjected to a sequential regimen of inorganic fertilizers, organic materials such as cow dung or chicken excrement, or a combination of both.

It has been discovered that a mixture of 750 kg of groundnut oil cake, 200 kg of cow dung, and 50 kg/ha of single super phosphate is very effective in producing the desired plankton. As mentioned, a thick paste from half of the amounts is spread throughout the nursery two to three days before stocking (Basudha *et al.*, 2017).

## Stocking

After emerging from their eggs three days after hatching, the recently hatched fry are relocated to specialized nurseries. It is best to carry out this relocation in the morning, affording them the opportunity to acclimate gradually to their unfamiliar pond surroundings. In earthen nurseries, it is suggested to maintain a spawn density ranging from 3 to 5 million per hectare. However, in cement cisterns, higher densities, such as 10 to 20 million per hectare, can be effectively employed.

During the nursery phase, it is generally advised to prioritize a monoculture approach, particularly focusing on carp species (Jayasankar *et al.*, 2018).

**Stocking of spawn in the nursery pond**

*(http://agropedia.iitk.ac.in/content/stocking-and-post-stocking-management-nursery-pond)*

## Post-stocking Pond Management

During the 15-day culture period, which was previously discussed, the phase fertilization is carried out in a couple of split doses. During the initial five-day period, an additional diet consisting of finely powdered groundnut oil cake and rice bran is provided at a quantity of 6 kilograms per million units. Subsequently, for the remainder of the duration, two equivalent portions of 12 kilograms per million units per day are administered. days. When scientific methods are used, the fry reaches the desired size of 20–25 mm with a survival rate of 40–60% after 15 days of rearing. Due to the 15-day limit on the nursery-rearing period, the same nursery can be used for multiple cropping, at the very least for raising 2-3 crops in the case of earthen ponds and 4-5 crops in the case of cement cisterns (Kumar *et al.*, 2021).

## 2. Fry- Fingerlings Rearing Pond Management

Ponds that are relatively larger than nurseries, typically ranging up to an area of 0.2 hectares, are commonly used for rearing purposes, specifically for raising fry to fingerlings. Several management practices are involved in maintaining these ponds, which can be summarized as follows (Ngugi *et al.*,2011).

## Pre-stocking Pond Preparation

The procedures employed for pond preparation prior to stocking, including the eradication of aquatic vegetation and the removal of predatory and undesirable organisms, closely mirror the techniques discussed in the context of managing nursery ponds. The quantity of organic manures and inorganic fertilizers introduced into the ponds is contingent upon the type of fish toxin employed. Typically, cow dung is applied at a rate of 10 tonnes per hectare, but if mahua oil cake serves as the fish toxin, the application is reduced to 5 tonnes per hectare of cow dung. Of the prescribed dose, one-third is administered as an initial base dose 15 days prior to stocking, while the remaining two-thirds are divided and applied at two-week intervals. Additionally, it is advisable to apply urea and single superphosphate at rates of 200 kg and 300 kg per hectare per year, respectively, at the same biweekly intervals, as sources of inorganic fertilization. (Boyd and C.E 2018).

## Stocking of Fry

The recommended stocking density for fry in rearing ponds typically falls within the range of 0.1 to 0.3 million per hectare. The specific stocking rate is primarily influenced by two key factors: the overall productivity of the pond and the particular management strategies that are planned to be implemented. During the nursery phase, the focus remains on cultivating a single species (monoculture), while the rearing phase embraces a more diverse approach, incorporating polyculture techniques that involve the simultaneous cultivation of various carp species. This approach is akin to the methods employed in the grow-out production phase (Sumagaysay,2007).

## Post-stocking Pond Management

Additionally to blending groundnut oil cake and rice bran in a 1:1 weight proportion, supplementary nourishment might encompass unconventional components. To foster fingerlings (Nandeesha *et al.*, 2013), a feeding rate of 5–10% was subsequently employed. For common or grass carp stocking, duckweeds or aquatic macrophytes like Wolffia, Lemna, and Spirodela should be dispensed in precise quantities. It is also advisable to contemplate other management strategies such as periodic fertilization. The water levels should be maintained at approximately 1.5 meters in depth. When employing advanced rearing methodologies, fingerlings achieve an adult size of 80–100 mm/8–10 g with a survival rate ranging from 70% to 90%.

# Grow Out Carp Culture

## 3. Carp polyculture

Carp polyculture in India has the potential to attain annual production rates ranging from 1 to 3 tonnes per hectare. This can be accomplished by effectively harnessing organic waste materials alongside both organic and inorganic fertilizers. However, a more scientifically driven approach can push these production levels even further, reaching 4 to 8 tonnes per hectare per year. This scientific approach involves a strategic amalgamation of nutritionally balanced feeds and appropriate fertilizers.

To optimize the management of carp polyculture, it is recommended to utilize ponds spanning from 0.4 to 1.0 hectares in size, with a water depth of 2 to 3 meters. These specifications are considered ideal for ensuring efficient and sustainable operations within this aquaculture system. Carp polyculture management encompasses various key stages, including pre-stocking, stocking, and post-stocking activities. These stages involve meticulous environmental and biological interventions to create a conducive ecosystem for carp growth and development (Jayasankar *et al.*, 2018).

### Pre-stocking Pond Preparation

Ponds can be made profitable by clearing them of weeds and predators and producing enough natural food sources. The essential components of preliminary pond management encompass the mitigation of aquatic vegetation and other unwanted organisms, alongside the improvement of soil and water conditions. Nursery management has already covered the specifics of controlling weedy and predatory fishes (Jadhav 2009).

### Stocking of Ponds

After the process of fertilization, the pond is carefully prepared. The appropriate-sized seeds are introduced once the sources have acclimated to their new surroundings. In the case of fingerlings destined for grow-out culture ponds, it is suggested that a size exceeding 100 mm is ideal for stocking. This size selection aims to achieve optimal survival rates and growth results. A 50-100 g size range is preferred for intensive polyculture ponds. This size field has yielded survival rates of over 90% and promoted better growth. In carp polyculture, a standard stocking rate of 5,000 fingerlings per hectare is typically maintained to achieve a production target of 3-5 tons per year. However, higher stocking densities have been utilized to increase production levels. Stocking densities within the range of 8,000 to 10,000 fingerlings per hectare have been utilized to achieve annual production rates ranging from 5 to 8 tons per hectare. Additionally, to reach even greater fish production targets of 10 to 15 tons per hectare per year, stocking densities in the range of 15,000 to 25,000 fingerlings per hectare have proven to be effective.

In the context of carp culture in India, a combination of six species has been recognized as the ideal composition. These various species encompass Catla, Silver Carp, Rohu, Grass Carp, Mrigal, and Common Carp. Each of these species exhibits distinct feeding behaviors: Catla and Silver Carp primarily feed near the water's surface, Rohu prefers to feed in the water column, Grass Carp primarily feed on macro-vegetation, and Mrigal and Common Carp are primarily bottom-dwellers when it comes to their feeding habits (Singh, 2012). The typical distribution of these species within a pond commonly adheres to a pattern where approximately 30-40% are surface feeders, 30-35% are mid-water feeders, and the remaining 30-40% are categorized as bottom feeders. However, the specific distribution may be adjusted based on the productivity of the pond.

**Optimum basic parameters of water**

| Physical parameters | Brood pond |
|---|---|
| Temperature Colour Greenish | 20-35 |
| Turbity in cm | 8-20 |
| **Chemical parameters** | |
| pH | 7.2-7.5 |
| $DCO_2$ | L4.0-12.0 |
| Total alkalinity | 80-150 |
| Ammoniacal Nitrogen $NH_4N$ (mg/1) | 0.2-0.5 |
| Nitrate Nitrogen $NO_2N$ (mg/l) | 0.2-0.2 |
| Nitrite Nitrogen $NO_2$ - N (mg/1) | <0.014 |
| Phosphorus $P_2$ $O_5$ (mg/1) | 0.01-0.5 |
| Iron (mg/g) | 0.05 - 0.02 |
| Manganese (Admissible range) (mg/1) | 0.01 - 0.04 |

*Source: Post-stocking Pond Management*

## Fertilization

In preparation for stocking, the recommended procedure involves the application of 20-25% of organic manures as a basal dose, ideally two weeks prior to stocking. The remaining portion of organic manure should be divided into equal instalments and applied on a bimonthly basis. Additionally, other readily available organic manures such as poultry litter, pig or cow dung, duck droppings, and domestic sewage can be effectively utilized. It's noteworthy that Azolla, a nitrogen-fixing macrophyte, has gained approval as a valuable biofertilizer for aquaculture purposes. The recommended usage rate for Azolla stands at 40 tonnes per hectare per year. This application rate is preferred because Azolla provides a comprehensive range of essential nutrients essential for the successful management of intensive carp culture.

This includes 100 kilograms of nitrogen, 25 kilograms of phosphorus, 90 kilograms of potassium, and 1,500 kilograms of organic matter.

| **Fertilizer requirements** | | | |
|---|---|---|---|
| | **Low** | **Med** | **High** |
| Available Nitrogen mg/100g soil | <20 | 25-50 | >50 |
| Available Phosphorus mg/100g soil | <3 | 3-6 | >6 |
| Organic carbon (%) | <0.5 | 0.5-1.5 | >1.5 |
| **Inorganic Fertilizer: (In splitted dose as per requirement)** | | | |
| | **Low** | **Med** | **High** |
| Nitrogen (kg/ha/yr) | 200 | 150 | 100 |
| Phosphorous kg/ha/yr (Single Super Phosphate) | 100 | 75 | 50 |
| Note: 1 kg Urea | | 0.46 kg Nitrogen | |
| 1 kg SSP | | 0.16 kg Phosphorus | |
| **Organic Manure** | | | |
| **Raw Cowdung & Liming** | | **@ 50 - 10 t/ha - Base manure** | |
| **Soil pH** | | **Dose of Lime ($CaCO_3$) kg/ha** | |
| 4.0-4.5 | | 1000 | |
| 4.5-5.5 | | 700 | |
| 5.5 - 6.5 | | 500 | |
| 6.5-7.5 | | 200 | |

*Source: (Handbook of fisheries and aquaculture)*

## Supplementary feeding

In the practice of carp polyculture, supplemental nutrition commonly consists of a blend comprising groundnut or mustard oil cake in combination with rice bran. As fish farming intensifies, diverse elements sourced from both plant and animal proteins are being introduced into these feeds. To maintain cohesion and enhance water stability while minimizing waste, these nutritional components are often pelletized. Additionally, designated sections of the pond are allocated for the cultivation of preferred aquatic plants such as Hydrilla, Najas, Ceratophyllum, duckweeds, and others, which are subsequently provided as sustenance for grass carp. Furthermore, alternative food sources like banana leaves, various land-based fodder, assorted grasses, and vegetable scraps can also be employed in the feeding regimen. For optimal results, it is recommended to feed the fish twice daily (Mayekar *et al.*, 2022). For the first month, feeding is done at 5% of the initial biomass of the stocking material and then at a sliding scale from 3-1% in the following month.

**Feed specially designed for the nutritional needs of Indian Major Carps**

| **Ingredients (in kg)** | |
|---|---|
| GNOC | 70.00 |
| Rice bran | 28.40 |
| NaCl | 1.50 |
| Trace element | 0.10 |
| **Ingredients (in g)** | |
| $FeSO_4$ | 50.00g |
| $CuSO_4$ | 8.00g |
| ZnO | 6.70g |
| $MnSO_4$ | 15.40g |
| Potassium iodide ( Kl) | 4.20 |
| $CaCl_2$ | 2.00g |
| $CaCO_3$ | 13.70g |
| Vitamins supplement per 100 Kg of feed | |
| Vit C | 200000 IU/10g |
| Vit E | 3000 IU/3g |

## Water Exchange and Aeration

Ponds' dissolved oxygen concentration can be artificially increased through aeration, which is especially important in intensive cultures with higher stocking densities. Another significant activity that is essential to intensive aquaculture is water exchange. The water quality degrades due to the constant buildup of metabolites and decayed unutilized feed, which slows down fish species' growth and frequently causes disease outbreaks. Consequently, it is crucial to periodically replenish a specific volume of water, particularly as the culture period advances and farming practices become increasingly demanding (Roy *et al.*, 2021).

### Harvesting

Harvesting fish typically occurs after ten months to one year of cultivation. However, fish that have reached a size suitable for the market can be collected periodically. This practice helps alleviate the density pressure within the pond, allowing ample room for the growth of other fish species. Regularly harvesting marketable-sized fish creates and maintains space to support the development of additional fish within the pond.

## References

Basavaraja, N. and Basavaraja, N. (2007). 7.11 Freshwater fish seed resources in India. *Assessment of freshwater fish seed resources for sustainable aquaculture*, 24(501), 267.

Basudha, C., Devi, N. S. and Sinthoileima, C. (2017). Bangana dero: A potential indigenous fish species for diversification carp culture in northeast India for sustainable aquaculture. *Aquaculture Asia Magazine*, 21(1), 19-23.

Boyd, C. E. (2018). Aquaculture pond fertilization. *CABI Reviews*, (2018), 1-12.

Hussan, A., Chakrabarti, P. P., Sundaray, J. K., Das, A., Mohapatra, B. C., and Ananth, P. N. (2018). Status and future of aquaculture development in Mizoram, India. *International Journal of Fisheries and Aquatic Studies*, 6(4), 42-48.

Jadhav, U. (2009). *Aquaculture technology and environment*. PHI Learning Pvt. Ltd..

Jayasankar, P., Mohanta, K. N.,and Ferosekhan, S. (2018). Freshwater aquaculture in India In: Aquaculture in India. Tripathi, S.D., Lakra W.S., Chadha, N.K.Edn. 23-49.

Kumar, A., Kaur, S. and Kaur, H. (2021). The emergence of black fungus and the covid-19 pandemic. *Innovative and current advances in agriculture and allied sciences*.

Mayekar, T. S., Sreekanth, G. B., Patil, A. S., and Paramesha, V. (2022). Entrepreneurship opportunities in Aquaculture for Livelihood Improvement: A present status and future prospectus. In: Ecosystem service analysis: concepts and applications in diversified coconut and arecanut gardens (pp.116-132) Publisher: ICAR Central Coastal Agricultural Research Institute, Goa

Nandeesha, M. C., Sentilkumar, V., and Antony Jesu Prabhu, P. (2013). Feed management of major carps in India, with special reference to practices adopted in Tamil Nadu. *On-farm feeding and feed management in aquaculture. FAO Fisheries and Aquaculture Technical Paper*, 583, 433-462.

Ngugi, R. W., Bowman, J. R and Omolo, B. O. (2011). Kenya. Remittance Markets in Africa, edited by Sanket Mohapatra and Dilip Ratha. Washington DC: The International Bank for Reconstruction and Development/The World Bank.

Roy, S. M., Machavaram, R., Pareek, C. M. and Mal, B. C. (2021). Diversified aeration facilities for effective aquaculture systems—A comprehensive review. *Aquaculture International*, 29, 1181-1217.

Singh, A. K. and Lakra, W. S. (2012). Culture of *Pangasianodon hypophthalmus* into India: impacts and present scenario. *Pakistan Journal of Biological Sciences*, 15(1), 19.

Sumagaysay-Chavoso, N. S. (2007). Analysis of feeds and fertilizers for sustainable aquaculture development in the Philippines. *FAO Fisheries Technical Paper*, 497, 269.

Wang, H., Amesa, S., Endebu, M., Tirfessa, G., Zhong-Yi, Z. and Zhi-Gang, W. (2018). Preliminary study on the construction of three-in-one fishpond and its effect on aquaculture in Ethiopia. *Fisheries and Aquaculture Journal*, 9(2), 1-10.

CHAPTER

# 5

# Biology of Indian Major Carps and Exotic Carps

## Indian Majar Carps (IMC)

The Indian major carps, also known as the carps, whose origin is India or indigenous carps, refer to commercially important fish species in India. These fish species include Catla, Rahu, and Mrigal. They hold significant economic value and are widely recognized for their contribution to the fishing industry in India.

## Biology of Catla

The scientific name of Catla is *Catla catla*. This species is known for its remarkable growth rate, surpassing other major carp.

**Catla**

*(https://hutterstock.com/search)*

## Classification

- **Phylum** - Chordata
- **Group**- Pisces
- **Class**- Osteichthys
- **Order**- Cypriniformes
- **Family**- Cypriniidae
- **Genus**- *Catla*
- **Species**- *catla*

## Characters

- The fish in question resides in freshwater habitats.
- It possesses a notable characteristic of a large, upturned mouth with prominent lips.
- Its body exhibits an elongated and broad structure, lacking barbels.
- The head of this fish is relatively large, accompanied by round eyes.
- Its colouration features a blackish-grey shade on the upper side, while the sides display a silvery hue.
- The caudal fin of this fish is bilobed, meaning it is divided into two distinct lobes.

## Food and feeding habits

The catla is a surface feeder, primarily relying on food sources near the water's surface. During its spawn-to-fry stage, it predominantly feeds on phytoplankton, tiny plant-like organisms suspended in the water. As the catla grows from fingerling to the adult stage, its diet includes crustaceans and insects. It can be categorized as a zooplankton animal, with approximately 60% of its diet comprising zooplankton and 40% composed of phytoplankton (Lalit *et al.*, 2015).

## Spawning Season

The spawning season for catla in rivers occurs from July to September, taking place once a year during this period. This is when catla individuals engage in reproductive activities to produce offspring.

## Maturity

The age and size at which fishes attain maturity can vary depending on the climatic and ecological conditions they inhabit. In the case of catla, it reaches sexual maturity within approximately two years of its life cycle. Catla individuals can reproduce and participate in the spawning season at this stage.

## Growth

Catla, one of the fast-growing major carp species in India, has the potential to reach a length of 6 feet and weigh over 60 kg. In well-maintained nursery ponds, hatchlings of catla are stocked at a rate of 10-12.5 lakh per hectare and can grow to a length of 20-25 mm within 15 days.

According to (Sen 2014), catla typically grows to a size of 38-45 cm (900 gm) by the end of the first year, reaches 4-5 kg by the second year, and attains a weight of 6-7 kg by the end of the third year.

## Fecundity

A female catla weighing 5 kg was found to have a total of 4,00,000 ova (eggs). As per Jhingran's study in 1963, the fecundity of catla varies between 2,30,831 and 42,02,250, depending on the length and weight of the fish's ovaries.

## Distribution

Catla is a freshwater fish in various regions, including India, Pakistan, Burma, and Bangladesh. It has also been transplanted to other countries such as Sri Lanka and Israel.

## Economic Importance

Catla holds significant economic importance as it is considered an important food fish. It is extensively utilized in pond culture throughout the country. Furthermore, it plays a crucial role in composite fish culture, which involves the integration of multiple fish species in the same aquaculture system to enhance productivity and sustainability.

# Biology of Rohu

The scientific name of Rohu is *Labeo rohita*. Rohu also grows very fast but is relatively little slower than catla.

## Classification

- **Phylum -** Chordata
- **Group-** Pisces
- **Class-** Osteichthys
- **Order-** Cypriniformes
- **Family-** Cypriniidae
- **Genus-** *Labeo*
- **Species -** *rohita*

**Rohu**

*(https://hutterstock.com/search/rohu)*

## Characteristics

- The body of Rohu is elongated with a rounded abdomen.
- The colour of the body ranges from brownish-grey to black above.
- The scales are large and arranged reddish at the centre.
- The head of Rohu is prominent, with thick lips.
- The mouth is located below the terminal position.
- The caudal fin of Rohu is forked and lacks barbels.

## Food and Feeding

Rohu is a column feeder and primarily feeds on vegetable debris, microscopic plants, decayed matter, and mud.

### Spawning seasons

The spawning season of the Rohu fish is from June to August. In the river Ganga, it breeds from April to August, as mentioned by David in 1959.

## Maturity

Rohu fish attains maturity at the age of 2 years. Males usually mature earlier than females.

## Growth

In well-maintained nursery ponds, when hatchlings of Rohu fish are stocked at a rate of 10-12.5 lakhs per hectare, they grow to a length of 25-30 mm in just 15 days. According to Hora (1962), Rohu fish can grow to a length of 35-45 cm (900 grams) in one year, reaching 2-3 kg in the second year and 4-5 kg by the end of the third year. After the fourth year, the growth rate slows down significantly.

## Fecundity

The absolute fecundity of several Rohu fish specimens, ranging from 51-75 cm in length and weighing between 1.5 to 7.6 kg, was found to vary from 4,25,600 to 27,94,000 eggs.

## Distribution

The fish in question is primarily found in freshwater habitats. It is known to inhabit various regions, including Pakistan, Bangladesh, and Burma. Furthermore, this fish has been successfully transplanted to numerous countries due to its adaptability.

## Economic Importance

This particular fish species is highly advantageous due to its rapid growth rate. It is extensively utilized for aquaculture purposes, predominantly in ponds nationwide. Moreover, it holds significant value as a food fish, making it a vital resource.

# Biology of Mrigal

The Mrigal, scientifically known as *Cirrhinus mrigala,* is a fast-growing fish species closely related to the Rohu.

## Classification

**Mrigal**

*(https://hutterstock.com/search/ Mrigal)*

- **Phylum-** Chordata
- **Group-** Pisces
- **Class-** Osteichthys
- **Order-** Cypriniformes
- **Family-** Cypriniidae
- **Genus-** *Cirrhinus*
- **Species-** *mrigala*

## Characteristics

- The Mrigal has an elongated body with a small head.
- Its mouth is located at the terminal end.
- The body of the Mrigal displays a silvery dark grey colouration.
- The pectoral and anal fins are black.
- The caudal fin is sharply forked.

## Food & Feeding Habits

The Mrigal is a bottom feeder and an omnivorous animal. Its diet includes copepods, decayed plants, algae, animal matter, mud, and phytoplankton, among other things.

## Spawning Seasons

The spawning season for Mrigal typically occurs between June and early August.

## Maturity

Both male and female Mrigal reach maturity when they are two years old.

## Growth

When the fry are stocked at a density of 10 to 12.5 lakh (1 to 1.25 million) per hectare, they grow to a length of 25-30 mm within 15 days. Fingerlings have been reported to reach a length of 55-65 cm and a weight of 1-2.2 kg within one year. According to (Sen *et al.*, 2014), Mrigal can reach weights of 1.8 kg, 2.6 kg, and 4 kg by the end of the first, second, and third year, respectively.

## Fecundity

Fecundity refers to the reproductive capacity or potential of a species. In the case of the mrigal fish, the weight of a female mrigal is recorded to be 1.47 kg, and a total of 2,16,800 eggs were found.

## Distribution

The mrigal fish is distributed primarily in North Indian rivers. However, it can also be found in Pakistan, Bangladesh, and Burma.

## Economic Importance

From an economic standpoint, the mrigal fish holds significant importance. It is widely cultured as a carp species throughout India and is highly valued as a food fish.

# Biology of Grass Carp

The grass carp exhibits several distinctive characteristics. Its body is elongated and moderately compressed, while its head is broad with a short, round snout. The upper jaw is slightly longer than the lower jaw, and it lacks barbels. The dorsal fin rays are short, with the dorsal origin slightly in advance or opposite to the ventral origin. The scales on its body are of moderate size. Regarding colouration, the grass carp is dark grey on its upper side and silvery on its belly (Zhao *et al.*, 2020).

**Grass carp**

*(https://hutterstock.com/search)*
*(Grass carp)*

## Classification

- **Phylum-** Chordata
- **Group-** Pisces
- **Class-** Actinopterygii
- **Order-** Cypriniformes
- **Family-** Cypriniidae
- **Genus-** *Ctenopharyngodon*
- **Species-** *idella*

## Food and Feeding

Regarding food and feeding habits, early fry to fingerlings primarily feed on zooplankton. As they grow into the adult stage, they transition to a diet of macro vegetation, including species like *Hydrilla, Ceratophyllum, Lemna, Wolffia,* and the soft parts of various plants.

## Growth

The grass carp exhibits a fast growth rate. Fry stocked at a rate of 10-12.5 lakh per hectare can reach a length of 22-27 mm in just 15 days. When stocked together with silver carp, they can grow to 12 cm in 3 months. By the time they reach three years, they can attain a length of 73-86 cm (4-7 kg) in weight (Cudmore and Mandark 2004).

## Maturity

Regarding maturity, grass carp in temperate regions typically reach maturity between the 2$^{nd}$ and 3$^{rd}$ year when their weight is around 2 kg. In colder countries, maturity is achieved at age 5-7 when the fish weighs about 2.5 kg (Cudmore and Mandark 2004).

Regarding fecundity, research conducted by (Alikunhi *et al.*,1963) found that specimens weighing 4-7 kg produced 3,0800 to 6,18,000 eggs.

## Spawning

Grass carp spawning occurs exclusively in lotic environments, particularly during the southwest monsoon period from June to September. They spawn once a year.

## Distribution

The grass carp is native to south, central, and north China and the middle and lower sections of the River Amur in the USSR. It has been transplanted and cultured in numerous countries worldwide, including India. In countries like Japan and Taiwan, it has become well-established.

## Economic Importance

The grass carp holds significant economic importance globally. It is widely cultured for biological weed control and is an essential component of polyculture systems.

## Biology in Silver Carps

**Silver carp**

*(https://hutterstock.com/search/ Silver carp)*

The scientific name the silver carp *Hypophthalmichthys molitric.*

## Classification

- Phylum- Chordata
- Group- Pisces
- Class- Osteichthys
- Order- Cypriniformes
- Family- Cypriniidae
- Genus- *Hypophthalmichthys*
- Species- *molitrix*

## Characters

The body is elongated and slightly compressed, with a pointed head and a bluntly rounded snout. The lower jaw protrudes somewhat, and the eyes are small in size. The scales on the body are tiny. The origin of the dorsal fin is located behind the ventral fin, approximately midway between the tip of the snout and the base of the caudal fin.

## Food and Feeding Habit

During the larval stages, the silver carp feeds primarily on nanoplankton, and as it grows into early and late fry stages, it shifts its diet to zooplankton. Zooplankton remains the primary food source, with occasional consumption of phytoplankton. However, as the silver carp matures, phytoplankton becomes its primary food source (Williamson *et al.*, 2005).

## Growth

When hatchlings are stocked at a rate of 10-12.5 lakh/hac, they grow to a size of 25-30 mm within 15 days. In China, the species achieves the following weights at the end of each year: 150 grams in the first year, 900 grams in the second year, 1,800 grams in the third year, and 4,200 grams in the fourth year. In reservoirs, the growth pattern is slightly different, with weights reaching 900 grams, 2,000 grams, and 4,500 grams to 5,000 grams during the first, second, and third years, respectively (Williamson *et al.*, 2005).

## Maturity

Silver carp typically reach maturity within 2-3 years of life in temperate countries when their weight is approximately 2 kilograms or higher. However, it may take 5-6 years to mature in frigid countries, weighing around 2.5 kilograms.

## Fecundity

A study by (Alikunhi *et al.*,1963) revealed that the total fecundity of silver carp, ranging in weight from 3-8.5 kilograms, varied from 145,000 to 2,044,000 eggs, with an average of 171 eggs per gram of body weight.

## Spawning

The spawning process of this particular species occurs once a year through hypophysation, which takes place from June to August.

## Distribution

Originally native to the freshwaters of China and Russia, this species has been successfully transplanted and cultured in numerous countries across temperate and subtropical regions worldwide, including India.

## Economic Importance

This carp species holds significant economic importance in India due to its rapid growth rate. It is highly valued for composite fish culture, making it a preferred choice in aquaculture practices. Additionally, it is widely utilized for stocking reservoirs and tanks abundant in phytoplankton, contributing to such aquatic systems' overall productivity and ecological balance.

# Biology of Common Carp

The scientific name of Common carp is *Cyprinus carpio* (common carp are three varieties):

- *Cyprinus carpio* (Scale carp)
- *Cyprinus carpio* var *nudus* (leather carp)
- *Cyprinus carpio* var *specularies* (mirror carp)

**Common carp**

*(https://hutterstock.com/search/ common carp)*

## Classification

- **Phylum-** Chordata
- **Group-** Pisces

- **Class-** Osteichthys
- **Order-** Cypriniformes
- **Family-** Cypriniidae
- **Genus -** *Cyprinus*
- **Species -** *carpio*

## Characters

The first ray of this organism's dorsal fin and anal fin has undergone modification, transforming into spines. These spines possess internal serrations. Additionally, two pairs of barbels are present, and the scales on the body are loosely arranged, shiny, and relatively small in size. Notably, the mouth of this organism is capable of protrusion.

## Food and Feeding Habits

The larval stages of the common carp feed on nanoplankton, while the fry feed on zooplankton. As they grow from fingerlings to the adult stage, their diet shifts to include benthic fauna, tubifex worms, other insect larvae, decayed vegetable matter, and residue of phyto and zooplankton. Adult common carp are bottom feeders and omnivores, readily consuming artificial food. They collect food by taking in bottom mud, ingesting digestible particles, and then expelling the mud, making the water turbid (Dadebo *et al.*, 2015).

## Growth

In terms of growth, according to (Mohd *et al.*, 2022 ), fry stocked at a density of 1.6 million per hectare can reach a length of 26 mm in 20 days with a 50% survival rate. In various ponds in India, fingerlings can grow to a weight of 1 kg within one year, assuming artificial feeding and a stocking rate of 2500 individuals per hectare.

## Maturity

The time common carp reach sexual maturity varies depending on climatic conditions. In the USSR, maturity is attained between 4-5 years. In the temperate zones of Europe, males reach maturity at four years and females at five years. In China and Japan, adulthood is reached at 2-3 years. In India, specifically in Cuttack, the Chinese strain of common carp matures at 6-8 months, with males maturing earlier (Alikunhi, 1967).

## Fecundity

The absolute fecundity of the common carp species is reported to range from 39,400 to 1,659,000 eggs. It is challenging to compare the fecundity of common carp from different regions, as it breeds 4 or 5 times a year in tropical countries and only once a year in temperate countries.

## Spawning

Regarding spawning, in temperate countries, common carp spawn once a year during spring, while in tropical waters, they can spawn throughout the year.

## Distribution

The fish is widely distributed, inhabiting freshwater rivers and ponds in China and Russia. It has also been successfully transplanted and cultured in various regions across the globe, including temperate and tropical areas such as India.

## Economic Importance

The largest cultivable and domesticated fish all over the world.

## References

Alikunhi, K. H. (1967). An account of the post-larval development, moulting and growth of the common stomatopods of the Madras coast. *In: Proceedings of Symposium on Crustacea, Marine Biological Association of India, part-II* (pp. 824-939).

Alikunhi, K. H., Sukumaran, K. K and Parameswaran, S. (1963). Induced spawning of Chinese carps, *Ctenopharyngodon idellus* and *Hypophthalmichthys molitrix* in ponds at Cuttack, India. *Proceeding Indo-Pacific Fishery Commission,* 10(2), 181-204.

Cudmore, B. M. N. E and Mandrak, N. E. (2004). Biological synopsis of grass carp (Ctenopharyngodon idella). *Canadian manuscript report of fisheries and Aquatic Sciences,* 2705(7), 1-44.

Dadebo, E., Eyayu, A., Sorsa, S and Tilahun, G. (2015). Food and feeding habits of the common carp (*Cyprinus carpio* L. 1758)(Pisces: Cyprinidae) in Lake Koka, Ethiopia. *Momona Ethiopian Journal of Science,* 7(1), 16-31.

Hora, S. L. (1962). Handbook of fish culture in the Indopacific region. *FAO Fish Biol.* Tech. Pap., 14, 1-203.

Lalit, K., Sharma, B. K., Sharma, S. K., Upadhyay, B. and Mishra, V. (2015). Food and feeding habits of *Catla catla* (Hamilton) from lake Udaisagar, Udaipur. *International Journal of Fauna and Biological Studies*, 2(5), 06-08.

Mohd, I., Bhat, F. A., Balkhi, M. H., Shah, T. H., Bhat, B. A. and Wali, A. (2022). Monthly variation in gonadosomatic index of *Cyprinus carpio* var. communis in wild and captive conditions of Kashmir Himalaya.

Natarajan, A. V. and Jhingran, A. G. (1963). On the biology of *Catla catla* (Ham.) from the river Jamuna. In *Proc. Nat. Inst. Sci. India* (Vol. 29, pp. 326-355).

Sen, J. M. (2014). Fresh water fish species diversity in Matatila reservoir Bundelkhand region, district Jhansi UP, India. *Current Agriculture Research Journal*, 2(2), 94-97.

Williamson, C. J. and Garvey, J. E. (2005). Growth, fecundity, and diets of newly established silver carp in the middle Mississippi River. *Transactions of the American Fisheries Society*, 134(6), 1423-1430.

Zhao, Y., Zhang, L., Wang, C. and Xie, C. (2020). Biology and ecology of grass carp in China: A review and synthesis. *North American Journal of Fisheries Management*, 40(6), 1379-1399.

CHAPTER

# 6

# Commercially Important Fish Culture

## Trout Culture

Trout is a highly sought-after fish that serves two primary purposes: cultivated for human consumption or released into natural waters for sport fishing. This captivating species is renowned for its vigorous fighting spirit and ability to yield exceptionally high-quality meat, making it a favourite among fishing enthusiasts worldwide. Over the years, trout have been bred and introduced to various water bodies across the globe. Commercial trout farming has been a long-standing practice in the United States, while Europe has been nurturing its trout culture for over four centuries. Trout are primarily cold-water inhabitants, thriving in brooks, lakes, ponds, rivers, and streams. In the Indian subcontinent, you can come across trout inhabiting various regions, including Kashmir, Himachal Pradesh, Uttar Pradesh, the Nilgiris, the Kodai Hills, and the Munnar High Range. Among the numerous trout species raised, three are particularly prevalent: the rainbow trout (*Oncorhynchus mykiss*), also known as Salmo gairdneri, the brook trout (*Salvelinus fontinalis*), and the European brown trout (*Salmo trutta*). Trout possess distinct characteristics that set them apart from other fish species. They have reduced gills and a sleek, streamlined body, enabling them to navigate swiftly through their aquatic habitat.

Furthermore, trouts exhibit remarkable adaptability, thriving in environments with high oxygen levels and enduring freezing temperatures. Their strong locomotion abilities allow them to cling to and burrow within their surroundings. Trout feed using their rasping lips, enabling them to gather food from pebbles, rocks, and other surfaces. This unique feeding mechanism allows trout to extract nourishment from their immediate environment, sustaining their growth and overall well-being. Overall, trout's appeal as a delectable food source and a prized catch for sport fishing enthusiasts has led to their widespread cultivation and introduction in various regions worldwide. With their captivating attributes, endurance, and remarkable feeding habits, trouts continue to enchant both anglers and connoisseurs of fine seafood (Sigler,2016).

## Spawning

The spawning season of three trout species, namely *S. gairdneri, S. trutta,* and *S. fontinalis,* varies from September to February, October to December, and October to January, respectively. These particular types of fish have a preference for gravelly surfaces when it comes to safeguarding their eggs, as these eggs tend to stick to the gravel and other debris. Trout, for instance, create nests and lay their eggs in the beds of streams. However, aquaculturists resort to artificial fertilization due to lower hatching rates resulting from streambed fertilization. They utilize methods such as adjusting the duration of daylight exposure and controlling the temperature of the water to stimulate the development of reproductive organs, thus guaranteeing a continuous production of juvenile fish throughout the year (Stewart, 1980).

Trout, in the process of their life cycle, are commonly captured when they reach or approach maturity as they swim upstream. These caught trout are subsequently nurtured in ponds until they attain full maturity. To facilitate their growth, brood fish are introduced into small ponds where water flows continuously, and measures are taken to prevent them from leaping out by covering the ponds with netting. It is worth noting that the milt (sperm) of a solitary male trout can fertilize eggs from two different females. This biological characteristic enables a higher stocking ratio of female trout with a relatively lower number of male trout.

During the breeding season, trout exhibit sexual dimorphism. Male trout become more colourful and develop a hooked beak, while female trout develop more enormous bellies and reddish genital papilla. When mature, gentle pressure on the abdominal vent releases milt or eggs. Females stripped for egg collection. Eggs in black enamel/plastic container. Add male milt, mix with quill feather for fertilization. Water added post-mixing for egg swelling (no pre-mixing to keep sperm motility). Enhance survival: saline solution with 10L fresh water, 90g common salt, 2g potassium chloride, 3g calcium chloride. Fertilized eggs turn green, "green eggs," moved to hatcheries. Pre-transfer, remove foreign particles and dead eggs for ideal development (Abraham *et al.*, 2010).

## Transportation of Trout Eggs

During the transportation process, the fertilized and hardened eggs of trout undergo specific packaging and temperature maintenance measures. The eggs, which have undergone a 24-hour hardening process, are meticulously arranged within cardboard containers with dimensions of 20 x 30 x 20 centimeters. To ensure proper insulation and protection, the interior of the cardboard box is furnished with styrofoam padding.

Two damp sponges or cotton pads are strategically placed inside the container to establish an ideal setting for the eggs during transit. One sponge is positioned at the bottom, while the other is situated at the top. Permeable polyethylene bags holding around 4,000 eggs are meticulously arranged between these moist sponges and cotton pads. To maintain a low temperature within the carton and ensure the eggs remain in optimal conditions, a polyethene bag filled with 1 kg of broken ice is placed above the upper pad. This ice helps to regulate and lower the temperature within the carton, preventing any potential harm to the eggs. Once all the necessary arrangements have been made, these cardboard cartons containing the trout eggs are ready for transportation to various destinations (Ayyappan and Jena 2003).

## Hatchery Techniques

Trout hatcheries use concrete troughs, trays, incubators, or jars to incubate trout eggs. Adequate water circulation and effective filtration play a pivotal role, in addition to clear freshwater devoid of sediment, in establishing an appropriate habitat. In the past, containers were frequently employed for the hatching process, but nowadays, vertical flow incubators have become the favored approach.

Vertical flow incubators have multiple trays stacked vertically. They are made of stainless steel, fibreglass, aluminium, wood, PVC, or plastic. The trays in question are equipped with either perforated zinc sheets, glass grills, or mesh cloth on their lower sections, enabling the free passage of water. These trays come in various sizes, ranging from dimensions as small as 180 cm x 30 cm x 10 cm to as large as 500 cm x 100 cm x 50 cm.

Within the incubator, you'll find trays that consist of both an upper egg basket and a lower perforated compartment. To initiate the incubation process, eggs are carefully placed within the upper basket. Water is then introduced into the tray, following a flow pattern that starts by moving upward through the egg basket and then cascading downward into the lower tray. This water circulation system helps create the optimal environment for egg incubation. This process continues throughout the incubator. The upward water flows through the eggs serve two important purposes (Woolford, 2019).

Firstly, it allows for increased aeration, ensuring a sufficient oxygen supply to the developing eggs. Secondly, the flow facilitates the removal of metabolites and waste products, maintaining a clean and healthy environment for the developing trout embryos. In summary, trout eggs are incubated in hatcheries using concrete troughs with horizontally arranged trays or incubators. Vertical flow incubators are commonly used, consisting of trays with perforated bottoms and upper egg baskets. Introducing water in an upward flow through the trays ensures adequate oxygenation and efficient waste removal, promoting the healthy development of trout eggs (Heasman and Savva, 2007).

Hatching jars are a valuable tool utilized for the incubation of trout eggs. These specialized containers facilitate the hatching process by providing an optimal egg environment. The typical composition of a hatching jar includes a galvanized screen with a mesh size measuring 0.5 mm. Additionally, there is a layer of gravel at the bottom, serving the purpose of eliminating undesired particles. Delicately, the eggs are positioned on the filter surface inside the jar. This setup facilitates the entry of water through the inlet, its upward passage through the filter and eggs, culminating in its discharge through the outlet. Handling the eggs carefully during hatching is crucial, as they are susceptible to external factors. Exposure to direct sunlight can be detrimental to the survival of newly fertilized eggs, potentially leading to their demise. Therefore, it is necessary to shield them from sunlight and provide a suitable environment for their development (Akhter *et al.*, 2021).

During incubation, ensuring that the water within the hatching jars remains in motion and possesses a high oxygen content is essential. These conditions are vital for the proper growth and development of the trout eggs. Ideally, incubation should occur within water temperatures ranging from 8°C to 12°C, providing an environment conducive to successful hatching. Once the eggs have hatched, the resulting hatchlings are typically kept within the hatching jars for a certain period. This allows for further monitoring and care before they are released into natural waters or utilized for stocking purposes. The young fish should be placed in containers or trays until they exhibit signs of increased activity and are able to feed on their own without assistance. In summary, hatching jars play a crucial role in the incubation and early stages of development for trout eggs. By providing an environment that safeguards against detrimental factors and ensures the necessary conditions for growth, these jars contribute to the successful hatching of trout and the subsequent release of fry into natural water bodies (Ngaboyera 2021).

## Culture of Trout

The young fish experience a specialized nurturing procedure that commences within compact rearing containers, even prior to complete absorption of their yolk sac. During this stage, they are introduced to artificial feeds. Once ready, they are transferred to nursery ponds, where they grow into advanced fry. These nursery ponds can be constructed from concrete or stone walls, with sizes ranging from 2.5 x 1 x 0.75 meters to 9 x 1 x 0.75 meters. A water flow of 100 liters per minute is maintained within the nursery pond to maintain proper conditions.

As the fry develops into advanced stages, they are moved to rearing ponds and raceways for further growth until adulthood. Rearing ponds are natural bodies of water, while raceways are fish ponds with running water. The raceways are typically constructed in a series on both sides of a stream or river. Inlets and outlets in the form of zinc plate screens are used to regulate water flow. The raceways should range from 20 to 100 square meters with a depth of 1.5 meters. Water is sourced from the stream or river, with a flow rate of 50 litres per second into the ponds. In the United States, circular and oval ponds are frequently employed for aquaculture purposes. The stocking density within these ponds might be constrained to achieve a yield of 5-10 kilograms per square meter.. However, with proper management, high production rates of up to 200 kilograms per square meter are achievable in raceways ( Leitritz and Lewis 1980).

The cage culture of trout is a commonly practised method in fish farming. In a recent experiment, fingerlings were introduced to cages at a stocking density of 1.4 $Kg/m^2$. These fingerlings were fed a diet of 3% of their body weight. Over two months, these trout grew significantly, reaching a weight range of 27-88 grams. The trout's diet consisted of a blend of cattle organs, specifically spleen, heart, and lung, along with a mix of marine or freshwater discarded fish. Although there are numerous commercial trout feeds accessible in the market, the usual feeding schedule for the trout usually involves offering food 3 to 4 times daily. Different approaches are employed when it comes to feeding trout. One method involves spraying the feed onto the water's surface. Alternatively, the meal can be placed in bags or containers in specific pond areas. For demand feeders, a mechanism is set up wherein the feed is released into the water whenever a trout triggers a particular device. Automatic feeders can also be utilized for this purpose, for rearing trout fry, jars and drums are commonly used. These containers provide a suitable environment for nurturing the young trout until they mature.

## Air - Breathing fish culture

Murrels and catfishes are highly valued and sought-after species in the market due to their lean meat and minimal presence of intramuscular spines. These remarkable fish species exhibit a unique capability to extract oxygen from the air, owing to their specialized respiratory structures. This adaptation grants them the ability to endure extended periods outside of water. Various supplementary respiratory apparatuses can be observed in these fish, which include respiratory trees in *Clarias*, the labyrinthine organ in *Channa*, the air bladder in *Heteropneustes*, and the branchial chamber in other related species. These specialized organs empower them to intake and make use of atmospheric oxygen. One of the notable advantages of cultivating these air-breathing fish lies in their capacity to flourish in environments characterized by low levels of dissolved oxygen, such as shallow, polluted waters, abandoned ponds, and swamps. Their Culture requires relatively low risk and simple management techniques, primarily because of their capacity to survive outside of water. This characteristic makes them suitable for farming in areas where conventional fish species might struggle. The esteem and high market demand for murrels and catfishes and their hardy nature and unique respiratory adaptations make them valuable choices for aquaculture in such environments.

In India, numerous regions like Andhra Pradesh, Assam, Uttar Pradesh, Madhya Pradesh, Tamil Nadu, Karnataka, Maharashtra, Bihar, and Meghalaya boast some of the most significant natural habitats for air-breathing fish. These unique fish species are renowned for their carnivorous tendencies and remarkable adaptability to supplementary feeding. Due to the shallow waters in which these air breathers thrive, there is minimal energy wastage through respiration, which allows for the anticipation of good yields (Datta,2011).

## The culturable species of air-breathing fishes in India.

**Scientific name** - Vernacular name

*Channa straitus* - Big or striped murrel

*Channa punctatus* - Spotted murrel

*Channa marulius* - Giant murrel

*Clarias batrachus* - Magur

*Heteropneustes fossilis* - Singhi

*Anabas testudineus* - Koi or climbing perch

## Culturable Areas

Air-breathing fish culture requires shallow waters (50-75 cm deep) and can be conducted in ponds or cages in running water systems. Unlike conventional culturable carp species, air-breathing fish can thrive in waters unsuitable for carp culture. Shallow ponds benefit these fish by minimizing the energy expenditure required to reach the surface for oxygen intake.

## Seed Collection

Seeds of murrel, magur, and singhi are obtained from nature, despite successful induced breeding. Collecting seeds from nature is still the most reliable stocking method. Murrels reach maturity in two years and can breed year-round. Fry measuring between 2 to 4 centimeters in length can be harvested on an annual basis from rainfed ditches and shallow water bodies teeming with plentiful aquatic vegetation. The peak of their reproductive activity is typically witnessed just prior to the onset of the monsoon season (Pathak *et al.*,2022).

The young ones hatched from eggs from shoals, making them easy to collect. However, fingerlings may not exhibit shoaling behaviour. There are different types of murrel fry with distinct characteristics. The juvenile giant snakehead fish showcases a predominantly dark gray body featuring a striking orange-yellow stripe that stretches from its eye all the way to its caudal fin. In contrast, the young striped snake head fish boasts a crimson-hued body adorned with a reddish-golden streak, accompanied by a distinctive black stripe that originates at the eye and extends to the caudal fin. Lastly, the fledgling spotted snakehead fish sports a deep brown body marked by a shimmering golden-yellow lateral band and a mid-dorsal yellow line tracing its back.

In the realm of murrel culture, it is highly recommended to opt for stocking fingerlings over fry. This preference stems from the fact that murrel fry often display cannibalistic tendencies. Furthermore, the survival rate of fry resulting from induced breeding tends to be notably low, and nurturing their growth until they reach the fry stage poses a formidable challenge. Upon hatching, fry do not partake in any feeding activities during the initial two days. Consequently, it becomes imperative to initiate their training to accept supplementary nourishment, which is offered in separate ponds. This supplementary feed encompasses boiled eggs, silkworm pupae, minced trash fish, worms, yeast, and vitamin B. This feed is administered for approximately 15 days, constituting about 20% of their body weight. Within a month, the fry will transform and develop into fingerlings, measuring about 4-6 cm in length.

The breeding behaviour of catfish is fascinating. These fish breed twice yearly, and their peak breeding season coincides with the rainy season. During this time, distinct characteristics can be observed in different catfish fingerlings. Magur fingerlings can be easily distinguished by their longer dorsal fin and slate colour. On the contrary, Singhi fingerlings feature a brief dorsal fin and display a rosy hue. Meanwhile, Koi fingerlings are distinguishable due to a distinct dark spot located on their caudal peduncle and a greenish tint adorning the upper side of their bodies.

When it comes to the reproductive behavior of magur fish, they display fascinating patterns. These fish dig holes that are approximately 25 cm deep in the bund just beneath the water's surface. Following fertilization, the eggs attach themselves to aquatic grass and are diligently protected by the male magur. Remarkably, each of these holes can yield a substantial number of fry, with numbers ranging from 2,000 to 15,000 per individual hole. To collect these fry, specialized tools such as small, finely-meshed hand nets are employed. Once collected, the fry are transferred to nurseries, where they are carefully nurtured until they reach the fingerling stage, typically attaining a length of around 5 cm (Azadi 1985). Magur, also recognized as the walking catfish, is a widely distributed fish species, well-suited for successful cultivation in ponds to produce fry. To facilitate breeding, specialized compartments are created within the ponds. Wire screen enclosures measuring 1 x 1 meter are constructed along the edges of the pond bund. In the heart of every enclosure, there is an excavation featuring a 30 cm-wide opening, which is then meticulously furnished with a selection of aquatic flora. These compartments provide suitable conditions for catfish breeding and ensure the success of the fry production process.

In the process of raising both male and female magur fish, they are typically segregated into compartments. Within a span of ten days, it is possible to harvest approximately 5,000 fry from each of these compartments. Alternatively, one can choose to rear the male and female fish together in small ponds, with each pond having the capacity to accommodate up to 20,000 magur fish. To sustain their growth, these fish can be nourished with a diet consisting of filtered zooplankton or a combination of chopped fish meal and groundnut oil cake. Once the fry have been collected, they can be transferred to specialized nurseries and carefully nurtured for a period of approximately 15 days. It is worth emphasizing that the optimal time for harvesting young magur fish, also known as singhi, falls during the pre-winter season, coinciding with the rice paddy harvest and the subsequent exposure of low-lying fields. This particular period marks the peak season for the collection of magur fish seeds (Katiha *et al.*, 2003).

## Seed Transport

When transporting fry and fingerlings of air-breathing fishes, it is possible to do so without oxygen packing by utilizing either polythene drums or iron drums. It is important to note that during the Transportation, the carrier should allocate enough space for the fry or fingerlings to come up to the surface and breathe atmospheric air, which they are accustomed to doing. Moreover, the carrier should include several aquatic weeds such as *Vallisneria, Hydrilla, Myriophyllum,* and *Ceratophyllum.* These weeds serve the purpose of preventing the fish from leaping out during the transportation process. Nonetheless, when the journey distance is considerable, it is advisable to convey them using polyethylene containers brimming with oxygen to guarantee their welfare.

## Pond Management

Nurseries in aquaculture typically have an area of approximately 10 to 15 square meters and a water column depth of 50 centimetres. In the nurseries' pond, there exists a population ranging from 0.2 to 1.5 million fry per hectare. Prior to introducing new fish, a process of fertilization is undertaken, wherein raw cattle dung is used at a rate of 500 kilograms per hectare. Additionally, to manage and prevent insect infestations, a soap-oil emulsion is carefully administered into the nursery pond water. While it is essential to apply nursery management techniques to oversee the growth of magur and singhi fry and fingerlings collected from natural sources, the murrels must undergo a nurturing process in the nursery ponds before they are ready to be introduced into the habitat. Once the nursery management phase is completed, the fingerlings can be transferred to stocking ponds (Sarkar, 2000).

## Stocking

Uniform-sized fingerlings are carefully selected for stocking in the fishpond. Before stocking, the fingerlings undergo a disinfection process. The samples can undergo one of two treatments: either they are exposed to a 2% $KMNO_4$ solution for a duration of 5 minutes, or they can be submerged in a 200 ppm formalin solution for a period of 50 seconds. In case of any wounds on the fingerlings, a 0.3% acriflavine solution is applied for 5 minutes to facilitate healing (Asha and Anil 2021).

To prevent the escape of these fishes, measures should be taken to address their climbing or burrowing tendencies. It is recommended to strengthen the embankments of the pond by utilizing sturdy logs, timber, or alternatively, by enclosing them with bamboo canes or wire screens. The height of these barriers should be around 50 cm. Regarding stocking fingerlings in their culture system, higher numbers can be accommodated. In monoculture systems, a stocking density of 40,000 to 60,000 fingerlings per hectare is recommended for catfishes. In polyculture systems, a range of 20,000 to 30,000 fingerlings per hectare of catfishes is suitable. For giant murrels in monoculture systems, the stocking density is 15,000 fingerlings per hectare, whereas, for striped murrels, it is 20,000. As for spotted murrels, the recommended stocking density in monoculture systems is between 20,000 and 30,000 fingerlings per hectare. In a polyculture setup, it is advisable to introduce both striped and spotted murrels at a stocking density of 20,000 fingerlings per hectare, maintaining an equal 1:1 ratio between the two species. With appropriate care and management, it is feasible to engage in polyculture involving murrels, carp, and catfishes. Ensure carp safety when introducing air-breathing fish seeds. Only stock seeds when carps weigh 300 grams+ to avoid predation. This approach generates income from air-breathing fish and helps carp grow. Eliminating trash fish through the growing air-breather population benefits carp growth (Jayasankar, 2018).

## Feeding

Throughout the eight-month period of semi-intensive cultivation in still ponds, it is advisable to provide the designated daily ration to the air-breathing catfish population during the nighttime. This feeding regimen aims to enhance feed utilization and improve overall growth. To ensure a plentiful food supply for the growing air breathers, it is crucial to have a stocking pond that is abundant in animal food sources such as frog tadpoles and trash fish. If the available food source is inadequate, the option of growing tilapia in murrel and catfish ponds can be considered an alternative. Another viable option is using dried marine trash fish, which proves to be more cost-effective in fish culture.

**The ratio of feed in different months (catfish culture)**

| Period | Weight of feed Kg/day | Feed ratio |
|---|---|---|
| **(stocking density 50,000 / ha) trash fish: rice bran** | | |
| 1-2 month | 12 | 1:2 |
| 3-4 month | 24 | 3:3 |
| 5-6 month | 36 | 5:2 |
| 7-8 month | 48 | 1:3 |

The potential food choices for catfish encompass a range of options. These options consist of fish byproducts or leftovers from the slaughterhouse, or a combination of dried silkworm pupae, rice bran, and oil cake, with each component being used in equal proportions. Another cost-effective feed option for singhi is a mix of equal parts of oil cake, rice bran, and biogas slurry. Rice bran and poultry feed can be provided in a proportion of 3 parts rice bran to 1 part poultry feed. Additionally, for biogas slurry and rice bran, a ratio of 1 part biogas slurry to 2 parts rice bran is recommended. This should be administered at a rate equivalent to 5-8% of the animal's body weight.

The food can be evenly spread across the pond by either scattering small quantities from the embankment or positioning feeding baskets in close proximity to the shoreline. This ensures that all the fish in the pond can access the feed. In murrel ponds, it is possible to establish light traps designed to lure insects using illumination. These insects subsequently serve as a valuable protein-rich food source for murrels (Sarkar, 2000). Another cost-effective feeding option for trained murrel fingerlings involves providing them with dried marine trash fish that has been rehydrated in water. The following feeding schedule can be followed for providing this type of feed. Animal protein from sources such as slaughterhouse waste and silkworm pupae can also be used (Mohapatra *et al.*,2015).

| **The feed for murrels in different months** | |
|---|---|
| **Period Feed** | **Kg/day** |
| 1-2 month | 3.75 |
| 3-6 month | 11.25 |
| 5-6 month | 17.50 |
| 7-8 month | 26.00 |

## Growth and Production

Murrels and catfishes can attain marketable size in just a matter of months, depending on their specific type. With the right management techniques in place, giant murrels and striped murrels are capable of achieving a growth rate ranging from 1 to 2 kilograms annually and 0.75 kilograms annually, respectively. In sharp contrast, spotted murrels can reach a weight of 160 grams in a span of just 8 months. Comparatively, catfishes tend to exhibit a slower growth rate when compared to murrels. Magur catfish and singhi catfish, for instance, have the capacity to grow up to 0.2 kilograms and 0.1 kilograms, respectively. In terms of recommended feed conversion rate, these species generally thrive when provided with a ratio of approximately 2:1, which indicates that for every 2 units of feed they consume, they can convert it into additional body weight (Ninawe *et al.*, 2020).

When murrel fish are enriched with supplementary forage fish, they have the potential to achieve an annual yield of approximately 4 metric tonnes per hectare. In the case of Magur catfish, if they are provided with supplemental feeds consisting of dried trash fish and rice bran, they can produce an impressive

10 metric tonnes per hectare per year. Similarly, the Singhi catfish has the capacity to yield approximately 4.4 metric tonnes per hectare annually. In a polyculture system involving murrel and koi fish, where they are nourished with a combination of rice bran, mustard oil cake, and trash fish, the potential yield can reach up to 11.8 metric tonnes per hectare per year. When Magur and Singhi catfish are fed with a diet primarily consisting of rice bran and trash fish, they can collectively produce around 5 metric tonnes per hectare annually. (Chakraborty and Ray, 2019).

Mixed culture of 3 murrel species = 4 tons/ha/year with soaked and dried trash fish + fresh silkworm pupae as food. In intensive culture, magur catfish = 7 tons/ha in 5 months.

## Culture with Carps

By implementing a stocking density of 5000 fish per hectare, comprising a mix of Indian and Chinese carp alongside 1000 magur fingerlings, the carp yield amounts to 2518 kilograms per hectare per year, while magur production reaches 3711 kilograms per hectare per year. These results clearly indicate that adopting a polyculture approach is more financially rewarding, emphasizing the advantages of incorporating magur into the carp farming system. Additionally, when utilizing a stocking density of 20,000 magur per hectare, combined with the remaining carp population after a partial carp harvest, an outstanding production of 3.96 tonnes per hectare per year is attained. This remarkable achievement is made possible by employing a supplementary feed with a composition ratio of 50% rice bran, 30% fish meal, 17% groundnut oil cake, and 3% minerals.

Magur is suitable for composite fish culture alongside carp, serving as a viable alternative to common carp. Additionally, koi, singhi, and the highly prized carp makhana (*Euryale ferox*) are suitable species to Culture in conjunction with magur (Ayyappan and Jena 2003).

## Harvesting

During the summer, a reasonable time arises for collecting air-breathing fish from ponds. To commence this process, the pond is methodically drained, enabling easy access and retrieval of the desired aquatic specimens. Employing the aid of scoop nets or hand nets, the fish farmers efficiently harvest these air-breathers.

The significance of this endeavour lies in the substantial demand and subsequent market value that this particular fish species command. Consequently, cultivating these air-breathing creatures offers lucrative income opportunities to fish farmers, all achieved through straightforward and manageable management techniques.

## Sewage - Fed Fish Culture

Sewage represents a intricate concoction of turbid, soiled, and unpleasantly odorous liquid, deriving from the drainage systems of our residences, including both toilets and kitchens. This mixture encompasses dissolved minerals, organic nutrients, and suspended solid particles. Proper sewage disposal has become a global challenge due to the rapid pace of urbanization. This issue is a consequence of demophora, which refers to unchecked human population growth. In recent years, sewage has become a significant pollutant, particularly in inland rivers. It serves as a breeding ground for numerous epidemics and poses a grave threat to soil and water ecosystems (Osbome *et al.*, 2009).

To address this problem, a sustainable approach to wastewater management should involve the utilization of sewage through a well-balanced system primarily focused on aquaculture. The concept revolves around reusing or recycling sewage residues in an ecologically friendly manner. Many countries worldwide have long recognized sewage effluent's potential to enhance the fertility of freshwater ponds.

In India, around 3.6 million cubic meters per day (equivalent to 800 million gallons per day) of sewage is produced. Urban areas contribute to approximately 30% of this total sewage production, amounting to 1.9 million cubic meters per day. Nevertheless, only 1.3 million cubic meters per day, which is approximately 20.4% of India's daily sewage output, undergoes treatment in these urban centers. Consequently, a significant portion, nearly 80% of the country's daily sewage production, remains untreated and untapped. Around 0.126 million tonnes of manure can be obtained from this untreated sewage in a day, totalling approximately 46 million tonnes annually. This manure is sufficient to cultivate a 0.1 million-hectare annual fish crop.

Sewage can also be utilized for the purpose of cultivating fish. In India, there exist more than 130 fish farms that rely on sewage as their primary water source, encompassing a total area of 12,000 hectares. A notable example can be found in the Vidyadhari sewage-fed fisheries located in proximity to Calcutta. Here, the local fishermen have efficiently harnessed the city's sewage disposal systems to support their fish cultivation endeavors. These fisheries have showcased remarkable results, boasting a fish yield of approximately 1,258 kilograms per hectare (Kg/ha). The rich nutrient content of sewage not only aids in fish growth but also acts as an additional water source, thereby contributing to an overall boost in fish production (Jana 1998).

## Composition of sewage

The makeup of sewage may exhibit considerable diversity, contingent on both the geographical area and the time of year. Water constitutes the prevailing constituent of sewage, making up roughly 99% of its overall composition. The remaining 1% consists of solid particles suspended in the water. In Indian urban areas, the typical composition of wastewater reveals specific concentrations of key components. On average, sewage in these regions contains 52 parts per million (ppm) of nitrogen, 16 ppm of phosphorus, 45 ppm of potassium, and 350 ppm of biodegradable organic matter. Within this organic matter, the organic carbon component falls within the range of 25 to 40 ppm, and it maintains a carbon-to-nitrogen ratio of approximately 1:3. Furthermore, sewage often contains elevated levels of salts originating from various heavy metals such as zinc (Zn), nickel (Ni), chromium (Cr), lead (Pb), and other contaminants. Frequently, these levels surpass the permissible limits established for sewage discharge. The organic constituents found in sewage primarily consist of proteins, carbohydrates, and fats. The proportions of these components can fluctuate based on the nutritional habits and dietary patterns of the population contributing to the sewage. Specifically, the carbohydrate content in sewage encompasses easily degradable substances like starch, sugar, and cellulose. (Barrera *et al.*,2011).

**Table 4.3 Ecological parameters of sewage water and freshwater ponds**

| Parameter | Sewage water pond | Freshwater pond |
|---|---|---|
| pH | 6.9 - 7.3 | >7.0 |
| Dissolved oxygen | 0-2 | 5.0 - 9.0 |
| $CO_2$ | 20 - 35 | 10-15 |
| $H_2$(ppm) | <2.0 | Nil |
| Specific Conductivity (Mohs) | 35 - 400 | 20 - 50 |
| BOD (ppm) | 100 - 300 | 1 - 15 |
| Free $NH_3$ (ppm | 20 | Nil |
| Albumenoid $NH_3$ (ppm | Nil - 16 | Nil |
| Nitrites (ppm) | 0.06 - 1 | Traces |
| Nitrates (ppm) | 0.01 - 30 | Traces |
| Phosphates (ppm) | 7-20 | Traces |

**Sewage water:** High BOD & OC, Suspended solids block sunlight, reduce photosynthesis. Strong: 1200 ppm, Medium: 720 ppm, Weak: 350 ppm total solids. Dissolved salts increase specific conductivity. Acidity from acid production renders water unfit for life when pH drops below 4. Sewage adds organic matter, leading to aerobic decomposition, reduced dissolved oxygen, and anoxic conditions. Anoxia induces the buildup of organic substances and the generation of harmful gases like hydrogen sulfide ($H_2S$), methane ($CH_4$), and carbon monoxide (CO), which undergo chemical reactions with water, resulting in the formation of acidic compounds. The presence of sewage has an immediate impact on aquatic life, leading to eutrophication, the rapid proliferation of phytoplankton, the development of algal blooms, and an increase in zoo plankton populations. It is crucial to take into consideration the levels of dissolved and suspended solids, organic carbon, nitrogen, and biological oxygen demand (BOD) concentrations when incorporating sewage in aquaculture practices (Silva, 2022).

### Microbiological characteristics

The presence of non-pathogenic bacteria in domestic sewage far outweighs the number of pathogenic bacteria. These non-pathogenic bacteria are not harmful and can even serve beneficial purposes. They consist of gut bacteria that are typically encountered within the general population, playing a role in the generation of waste. In raw sewage, the typical range of coliform bacteria load is between $10^8$ and $10^9$ most probable number (MPN) per 100 millilitres (Silva, 2022)

## Site Selection and Construction of Sewage-fed Fish Farm

The fish farm near urban area + domestic sewage nutrient recycling. The best location is close to the sewage treatment plant. A farm should be lower than a plant for easy sewage flow. Use a gravity-fed pipeline. A farm needs drainage facilities for water removal.

The sewage source, the cultural system, and the topography of the land all affect the fish farm's design. The remaining portion is used for dykes and other purposes while nearly 75% of the total area is transformed into ponds. Rectangular fish ponds are built with a maximum depth of 1.5m and a slope of 1:3 for the embankment. They range in size from 0.3 to 1 ha. Each pond needs to have adequate drainage systems.

The effluent is gathered and kept in a sump at the farm. The effluent is then piped through a distributing system and into the ponds. Additionally, plans have been made to connect the pipelines to a source of fresh water, allowing for emergency dilution if necessary (Mukherjee and Jana 2007).

### Sewage treatment

Sewage treatment is vital to eliminate harmful microbes, prevent anoxia, raise pH levels to alkaline, enhance photosynthesis, and reduce organic content.

It must be affordable and replicate the conditions of a natural freshwater pond. Sewage is treated through mechanical, chemical, and biological methods.

## Mechanical treatment

During phage treatment, mechanical treatment methods effectively eliminate solids and organic matter. This involves steps such as flowing, dilution, and sedimentation. To remove waste solids, sewage undergoes screening and straining. The remaining liquid and semisolid wastes are mixed with a diluent to form a suspension. Furthermore, the levels of multiple gases, such as hydrogen sulfide ($H_2S$), carbon dioxide ($CO_2$), carbon monoxide (CO), ammonia ($NH_3$), and methane ($CH_3$), are lowered to levels that fall below the typical or expected concentrations. Following the initial treatment process, the clear liquid effluent is segregated from the solid sludge, resulting in the creation of the supernatant (Gerba and Pepar, 2019).

## Chemical treatment

Many dissolved compounds, detrimental microorganisms, and offensive odors are efficiently eradicated through chemical processing. Various affordable substances such as precipitants, coagulants, chelating, disinfectants, deodorizing, and more are utilized to achieve this. Additionally, sewage water involves chlorine, bleaching powder, and copper sulfate treatment. This specific method is commonly referred to as secondary treatment. (Zhang *et al.*,2017).

## Biological treatment

In the realm of biological sewage treatment, a crucial focus is placed on fostering the growth of bacteria. These bacteria play a pivotal role in driving the oxidation of organic materials. As a result of this enzymatic activity, nitrogen oxides are produced, initiating the rapid proliferation of algae, notably the blue-green Microcystis. This flourishing population of algae serves as a protective shield against oxygen depletion in the water, as it elevates dissolved oxygen levels, diminishes the concentration of $CO_2$, and shifts the pH from acidic to alkaline. Additionally, the algal bloom works to decrease the presence of dissolved salts within the sewage water (Gerba and Pepar,2019).

## References


Abraham, T. J., Sil, S. K., & Vineetha, P. (2010). A comparative study of the aquaculture practices adopted by fish farmers in Andhra Pradesh and West Bengal. *Indian Journal of Fisheries*, 57(3), 41-48.

Akhter, F., Siddiquei, H. R., Alahi, M. E. E. and Mukhopadhyay, S. C. (2021). Recent advancement of the sensors for monitoring the water quality parameters in smart fisheries farming. *Computers*, 10(3), 26.

Ayyappan, S. and Jena, J. K. (2003). Grow-out production of carps in India. *Journal of Applied Aquaculture*, 13 (3-4), 251-282.

Ayyappan, S. and Jena, J. K. (2003). Grow-out production of carps in India. *Journal of Applied Aquaculture*, 13(3-4), 251-282.

Azadi, M. A. (1985). Fisheries information bulletin.

Barrera-Díaz, C., Martínez-Barrera, G., Gencel, O., Bernal-Martínez, L. A. and Brostow, W. (2011). Processed wastewater sludge for improvement of mechanical properties of concretes. *Journal of hazardous materials*, 192(1), 108-115.

Chakraborty, A. and Ray, A. (2019). Deep-water rice and indigenous fish co-cultivation to ensure rural livelihood, food security, and ecosystem health.

Datta, S. (2011). Inland Fisheries Resources of India. *CIFE, Kolkata Centre, Article in Inland Water Biology*.

Gerba, C. P. and Pepper, I. L. (2019). Municipal wastewater treatment. In *Environmental and pollution science* (pp. 393-418). Academic Press

Heasman, M. and Savva, N. (2007). *Manual for intensive hatchery production of abalone*. NSW Department of Primary Industries, Port Stephens Fisheries Centre.

Ignatiou, B., Asha, A., and Anil, M. K. Package of Aquaculture practices.(2021)

Jana, B. B. (1998). Sewage-fed aquaculture: the Calcutta model. *Ecological Engineering*, 11(1-4), 73-85.

Jayasankar, P. (2018). Present status of freshwater aquaculture in India-A review. *Indian Journal of Fisheries*, 65(4), 157-165.

Katiha, P. K., Jena, J. K. and Barik, N. K. (2003). Profile of key inland freshwater aquacultural technologies in India. *A Profile of People, Technologies and Policies in Fisheries Sector in India*, 59.

Leitritz, E. and Lewis, R. C. (1980). *Trout and salmon culture: hatchery methods* (Vol. 164). UCANR Publications.

Lu, S., Zhang, X. and Xue, Y. (2017). Application of calcium peroxide in water and soil treatment: a review. *Journal of hazardous materials*, 337, 163-177.

Mohapatra, B. C., Barik, N. K., Anantharaja, K., and Jayasankar, P. (2015). Installation and Operation of FRP Carp Hatchery.

Mukherjee, S. and Jana, B. B. (2007). Water quality affects SDH activity, protein content and RNA: DNA ratios in fish (*Catla catla, Labeo rohita* and *Oreochromis mossambicus*) raised in ponds of a sewage-fed fish farm. Aquaculture, 262(1), 105-119.

Ngaboyera, E. (2021). *Automated fish eggs incubator with smart cylinder* (Doctoral dissertation, College of science and Technology).

Ninawe, A. S., Dhanze, J. R., Dhanze, R. and Indulkar, S. T. (Eds.). (2020). *Fish Nutrition and Its Relevance To Human Health*. CRC Press.

Osborne, M. R., Champlain, D., Garrett, C., Hall, R., Hengst, G., McCarty, P. and Model, A. S. (2009). *Biological Wastewater Treatment and reactive nitrogen.*

Pathak, H., Mishra, J. P., and Mohapatra, T. (2022). Indian Agriculture after Independence. *Indian Council of Agricultural Research, New Delhi*, 110 (001), 426.

Sarkar, S. K. (2000). Efficacy of some non-conventional feed ingredients on the kinetics of food intake growth promotion and associated Biochemical changes in Indian major carps.

Sigler, J. W. (2016). *Fishes of the Great Basin: a natural history*. University of Nevada Press

Silva, M. R. (2022). *Natural and Tailored Magnesium-Based Materials for Adsorptive Nutrient Recovery and Catalytic Wastewater Treatment* (Doctoral dissertation, Lehigh University).

Stewart, L. (1980). A history of migratory salmon acclimatization experiments in parts of the Southern Hemisphere and the possible effects of oceanic currents and gyres upon their outcome. In *Advances in marine biology* (Vol. 17, pp. 397-466). Academic Press.

Woolford, J. (2019). World abalone fisheries and stock enhancement. *Fisheries Research and Development Cooperation. Nuffield Australia Project*, (1702).

CHAPTER

# Fish Food Organisms

7

## Introduction

In an aquatic ecosystem, aquaculture's primary resource of great value is the organisms that serve as food for fish. These microscopic organisms naturally occur in the marine environment, serving as the primary nutrition source for finfish and shellfish larvae. They vary in size, ranging from tiny microns to a few millimetres. The types of fish food organisms present in a particular water body depend on the productivity of that body of water. These organisms include protozoans, bacteria, phytoplankton, zooplankton, crustaceans, and worms. These organisms are tiny, making them suitable for the mouth size of various finfish and shellfish larvae. During their development from eggs to larvae, they rely on yolk as their source of nutrition. This transitional phase is a critical and delicate period as the larvae undergo changes in size and their digestive system adapts. The hindgut epithelial cells are responsible for a significant portion of protein digestion. To address this challenge, fish food organisms play a crucial role. The considerable expansion of finfish and shellfish aquaculture worldwide can be attributed to advancing mass production techniques for live feed. Live feeds can swim in the water column, ensuring a constant food supply for the larvae.

The term "plankton" originates from the Greek word "plankton," meaning "drifter." Plankton relies on water currents for their movement. They are classified into two main groups: phytoplankton and zooplankton. Plankton is crucial in aquatic ecosystems' biology as they nourish marine organisms. They form the foundation of the food pyramid, with zooplankton and phytoplankton serving as a significant link in the aquatic food web. Phytoplankton is particularly important in producing organic material, while zooplankton acts as a critical component of secondary production, connecting the producers (phytoplankton) with consumers (Yadav, 2015).

## Phytoplankton

Phytoplankton refers to microscopic, single-celled plant communities in freshwater and marine ecosystems. They can exist as solitary cells or colonies and range in size, with some reaching 500μm. Like other plants, most phytoplankton are autotrophic and contain chlorophyll pigment, allowing them to harness solar energy through photosynthesis (Findlay and King, 2001). Various factors influence phytoplankton growth, including sunlight, salinity, temperature, pH, tides, water currents, and nutrients such as nitrogen, phosphorus, carbon, silica, and iron. Phytoplankton is present in various water bodies, and their abundance significantly impacts the surrounding environment. These organisms serve as primary producers within the aquatic food web.

Additionally, phytoplankton plays a vital role in the carbon cycle, utilizing carbon during photosynthesis and releasing oxygen into the water and atmosphere. It is estimated that phytoplankton contributes approximately 50% of the oxygen released into the air. Phytoplankton also indicates environmental changes (Manickam *et al.*, 2012). The main types of phytoplankton include diatoms, dinoflagellates, coccolithophores (prymnesiophytes), cyanobacteria, and green algae (Gireesh *et al.*, 2015). Plankton, specifically phytoplankton, is a critical component of the trophic structure as it facilitates energy transfer to higher trophic levels. Phytoplankton accounts for about 95% of the total marine plant production, making them essential for marine ecosystems (Yadav, 2015). Many fish and shellfish larvae naturally feed on tiny phytoplanktonic and zooplanktonic organisms (Das *et al.*, 2012).

## Zooplankton

Zooplankton, which consists of microscopic organisms, holds significant importance in the aquatic ecosystem. They are susceptible and responsive to external disturbances, making them valuable indicators of environmental changes such as global warming or increased atmospheric carbon dioxide levels (Dai *et al.*, 2016). Zooplankton is an indicator for monitoring water quality, trophic status, and pollution levels. They also play a crucial role in indicating the presence or absence of specific fish species and determining population densities (Kamble *et al.*, 2013). Furthermore, zooplankton contributes significantly to the secondary production of the aquatic ecosystem and occupies an intermediate position in the food web, transferring energy from lower to higher trophic levels. Fish and other species consume zooplankton, which is the most important animal group in the marine ecosystem (Ansari and Khan, 2014). Ostracods, copepods, rotifers, and cladocerans make up the major zooplankton groups.

## Infusorians

Infusoria, or protozoa or ciliates, constitute a varied collection of microscopic organisms. The term "infusoria" comes from the historical method of soaking materials like hay at temperatures below boiling. Only the largest protozoans, classified as infusoria, are visible to the naked eye, while most species require a microscope for proper observation. An abundance of infusoria, often due to overfeeding, can be indicated by the presence of "green water" or cloudiness in aquariums.

Infusoria play a significant role in pisciculture by serving as an essential food source for fry. Their small size, ranging from 25 μm to 300 μm, makes them highly suitable as live food for recently hatched fry that has depleted their yolk sac. This group includes various ciliates and protozoans. Infusoria acquire nutrients through holozoic feeding, consuming bacteria, algae, and detritus. They reproduce through either binary fission or conjugation. In freshwater environments, species such as Paramecium and Stylonichia are commonly found, whereas marine species like *Euplotes* thrive in salinities ranging from 20 to 35 ppt. Stylonichia is a type of ciliate belonging to the stichotrich group. It is widely distributed in freshwater and soil, often found on filamentous algae, surface films, and among sediment particles. *Fabreasalina,* on the other hand, is a heterotrichous ciliate reported in various environments, including salt marshes and hypersaline lakes.

## Rotifer

Rotifers are tiny, multicellular animals that belong to the species Rotifera. They are commonly found in freshwater environments, although some species can also be found in marine habitats and damp soil. Rotifers are known for their unique characteristics and reproductive strategies (Stelzer& Snell, 2003).

Key features and characteristics of rotifers:

- **Size**: Rotifers are usually microscopic, ranging from about 50 micrometres to a few millimetres long. Despite their small size, they exhibit remarkable complexity in their anatomy and physiology.
- **Body Structure**: Rotifers are creatures with unusual bodies made up of a head, trunk, and foot. They use the corona, which is a crown of cilia on the head, for eating and moving about. The head also contains sensory organs. The trunk contains internal organs, and the foot helps attach the rotifer to surfaces.

- **Reproductive Strategies**: Rotifers exhibit various reproductive strategies, including sexual and asexual reproduction. Some species reproduce by parthenogenesis, where females produce offspring without fertilization. In other cases, sexual reproduction occurs with separate male and female individuals.
- **Feeding Habits**: Most rotifers are filter feeders, using their cilia to create currents that draw in food particles. They have a specialized structure called the mastax, consisting of jaws and muscles that help them grind food. Rotifers are opportunistic feeders and can consume various organic materials, including bacteria, algae, and small suspended particles.
- **Environmental Adaptability**: Rotifers are incredibly resilient and adapt to various environmental conditions. Some species can tolerate extreme temperatures, high salinity, and low oxygen levels. Their ability to enter a dormant state called a resting or desiccated state, allows them to survive unfavourable conditions such as drying ponds.
- **Ecological Importance**: Rotifers play crucial roles in aquatic ecosystems. As filter feeders, they help regulate microbial populations by consuming bacteria and algae. They also serve as a food source for many other organisms, including small fish and invertebrates.

## Culture of Rotifera

The culture of rotifers refers to the maintenance and propagation of these microscopic organisms in a controlled environment. Rotifer cultures are commonly established in research laboratories, aquaculture facilities, and educational institutions for various purposes, such as studying their biology, using them as live food for other organisms, or conducting toxicity tests Snell & Janssen, (2011).

### Steps involved in culturing rotifers:

- **Culturing Container**: A suitable container is required to house the rotifers. This can be a glass or plastic vessel with a wide opening and sufficient volume to hold the desired quantity of water and rotifers.
- **Water Source**: Rotifers are typically cultured in freshwater. The water used should be good quality, free from contaminants and pollutants. Tap water can be treated or conditioned to make it suitable for rotifer culture.
- **Nutrient Enrichment**: Rotifers require a supply of nutrients to thrive. Adding a nutrient solution to the water helps support their growth and

reproduction. Commonly used nutrient sources include commercial rotifer feeds, such as algae-based powders or liquid formulations.

- **Temperature and Light**: Rotifers thrive at specific temperature ranges depending on the species. Maintaining a suitable temperature, usually between 20-30°C, helps promote their growth. Adequate lighting conditions, such as a light source with a relevant spectrum, can also be provided to support photosynthetic algae growth, which serves as food for rotifers.
- **Aeration and Agitation**: Rotifers require oxygen for respiration. Providing aeration in the culture vessel through an air pump or gentle agitation ensures dissolved oxygen is available throughout the culture medium.
- **Feeding and Harvesting**: Rotifers primarily feed on microscopic algae. Adding algae to the culture medium provides a food source for rotifers. The algae can be grown separately and introduced into the rotifer culture as needed. Rotifers can be harvested using techniques such as sieving, sedimentation, or specialized rotifer harvesting devices.

## Cladocera

Cladocera is a group of small, aquatic crustaceans commonly known as water fleas or Cladocerans. They belong to the class Branchiopoda and are found in various freshwater habitats such as lakes, ponds, and rivers.

## Classification

- **Kingdom -** Animalia
- **Phylum -** Arthropoda
- **Subphylum -** Crustacea
- **Class -** Branchiopoda
- **Subclass -** Phyllopoda
- **Order –** Cladocera

The widely recognized genera in this group include *Daphnia* (freshwater fleas), which has received the most research attention, and *Moina*. Cladocerans are organisms that swim freely and tend to position themselves with the dorsal side facing upwards. The head is usually distinguished from the body by a pronounced groove, although this separation may sometimes be absent. The lead extends forward as a beak or rostrum, and on the forehead is a single compound eye formed by the fusion of two eyes. Additionally, most species possess a simple naupliar eye.

- **Kingdom**: Animalia
- **Phylum:** Arthropoda
- **Subphylum:** Crustacea
- **Class**: Branchiopoda
- **Order:** Cladocera
- **Family:** Moinidae
- **Genus:** *Moina*

There are various species of Moina, including *Moina affinis, M. australiensis, M. belli, M. brachiata, M. brachycephala, M. flexuosa, M. hartwigi, M. hutchinsoni, M. macrocopa, M. micrura, Moina minuta, M. mongolica, M. rectirostris, M. reticulata, M. salina,* and *M. tenuicornis*. Average adult Moina sizes range from 700 to 1,000 m, making them similar in size to adult rotifers but smaller than freshly hatched brine shrimp. Moina's extended lifetime makes it ideal for feeding young freshwater fish.

The primary food supply for the fry of several ornamental tropical fish species, moina micrura is cultivated specifically in ponds fertilised mostly with chicken dung or occasionally with pig manure, leading to a high survival rate of 95-99%.

## Key features and characteristics of Cladocera:

- **Body Structure:** Cladocerans have a distinct body structure characterized by a small, elongated body covered by a protective carapace or exoskeleton. The carapace is typically transparent and resembles a bivalve shell, enclosing most of the body. The body is divided into the head, thorax, and abdomen.
- **Appendages:** Cladocerans possess specialized appendages for various functions. The head bears antennae and compound eyes. The thoracic region features pairs of branched appendages known as thoracic limbs or swimming legs, which are used for locomotion and creating water currents for feeding. The abdomen may have specialized appendages for reproduction and swimming.
- **Reproduction:** Cladocerans exhibit different reproductive strategies, including sexual and asexual reproduction. Under favourable conditions, they reproduce rapidly through parthenogenesis, a process in which females produce offspring without fertilization.

## Reproductive cycle of Moina

*Moina* exhibits a reproductive cycle that encompasses both sexual and asexual phases. Typically, the population is dominated by reproducing females engaged in asexual reproduction. In ideal circumstances, Moina can begin mating as early as 4–7 days old and can produce 4–22 broods per female. Every 1.5–2.0 days, these broods are generated, and most females have 2–6 broods in their lifetime. Males are created under adverse environmental circumstances, causing sexual reproduction and the formation of latent eggs known as ephippia.

The transition from asexual to sexual reproduction in Moina populations is stimulated by a sudden decrease in food availability, resulting in an increased production of resting eggs. However, since fewer children are generated by resting eggs, it is advantageous to maintain a well-nourished population in the asexual method of reproduction. Cladocerans are filter feeders, primarily consuming suspended particles in the water. They use their thoracic limbs to create currents that bring in food particles, including algae, bacteria, and residue. Some species have modified feeding appendages known as "filter baskets" or "grazing baskets" that aid in capturing food.

Ecological Significance: Cladocerans are essential components of freshwater ecosystems. They are crucial in nutrient cycling and energy transfer within food webs. By consuming algae and bacteria, they regulate water quality and contribute to the control of algal blooms. Cladocerans also serve as a food source for many other organisms, including fish and insects.

## Culture of Cladocerans

The culture of Cladocerans, specifically the water flea Daphnia. Maintaining a healthy and productive cladoceran culture requires specific techniques and considerations (Hagiwara & Hanazato, 1999).

### Steps involved in culturing Cladocerans, particularly Daphnia:

- **Culturing Container:** Cladocerans are typically cultured in transparent containers such as glass or plastic vessels. These containers should provide enough space for the population to grow and have a wide open for easy access.
- **Water Source:** Cladocerans, being freshwater organisms, require a suitable water source. Ideally, use de-chlorinated tap or filtered pond water to provide the necessary environment. Ensure the water quality is good and oxygenated, free from contaminants.

- **Food Source:** Cladocerans feed on algae and other small organic particles. Culturing green algae separately and adding it as a food source to the Cladocerans culture is common. Microalgae such as Chlorella or Scenedesmus are commonly used. Other options include commercial algal preparations or yeast-based feeds.
- **Temperature and Light:** Cladocerans have specific temperature requirements for optimal growth and reproduction. Maintaining the culture at a suitable temperature, typically between 20-25°C, is recommended. Adequate lighting conditions, such as a light source with the appropriate spectrum and intensity, are crucial for supporting algal growth and providing a light-dark cycle for the Cladocerans.
- **Aeration and Agitation:** Cladocerans require a well-oxygenated environment. Aeration through an air pump or gentle agitation helps ensure sufficient dissolved oxygen levels throughout the culture vessel. Gentle agitation also helps prevent the settling of particles and promotes water circulation.
- **Monitoring and Maintenance:** Regular monitoring of the culture is essential. This includes observing population density and water quality parameters (e.g., temperature, pH, ammonia levels) and adjusting the feeding regime accordingly.

## Copepods

Copepods are common zooplankton both in freshwater and brackish water. Their advantages are a wide range of body sizes within and between species. They are natural food sources for many finfish and crustacean larvae and young. Copepodites and early-stage nauplii can be excellent starter prey for animals with small mouth gags. During their first few weeks of life, the majority of marine larvae in the wild eat copepod eggs and nauplii. Several larval fish must either be grown on copepod nauplii raised in a lab or collected in the wild instead of utilising rotifers as their primary meal. Copepod adults have a length of 0.5 to 5.0 mm. Six naupliar and six copepodite stages make up the larval stages.

- **Kingdom:** Animalia
- **Phylum:** Arthropoda
- **Subphylum:** Crustacea
- **Class:** Maxillopoda
- **Sub-class:** Copepoda

## Biology and life cycle

Most mature copepods typically have a length ranging from 1 to 5 mm. They possess a cylindrical-conical body shape with a more comprehensive front section. The cephalothorax, which contains the fused head and the first of the six thoracic segments, and the abdomen, which is smaller than the cephalothorax, may be distinguished as the two different portions of the body. The chair has a long, unbranched first antennae and a central naupliar eye. Planktonic copepods use their second maxillae to collect food particles. They consume bacteria and phytoplankton. Male copepods are typically smaller in size and less abundant than females. The eggs develop through five to six naupliar stages (moults) as nauplii before hatching into copepodites. Copepodite moults reach adulthood and end after five moults.

Copepods have a diapause stage in their development, enabling them to survive unfavourable environmental conditions, including freezing. The major diapause habitat is the sediment, while a limited number of diapausing organisms may continue to exist in the planktonic fraction, a state known as "active diapause." Harpacticoid copepods are more suited for intensive rearing because they are less sensitive to and more tolerant of harsh environmental changes than calanoids (e.g., salinity: 15-70 g/l; temperature: 17-30°C). Harpacticoids may also be fed a variety of food sources, including as microalgae, bacteria, detritus, and even manufactured diets, and they are more productive than calanoids. However, because copepods rely so significantly on the supplied meal, it is important to take into account the lipid and (n-3) HUFA composition of copepods.

## Culture techniques

A continuous production system for *Acartia tonsa*, a calanoid copepod, comprises three units: basis, growth, and harvest. *A. tonsa* can be produced by either raising it from resting eggs or isolating it from natural plankton samples. For the purpose of maintaining population stocks, the 200-liter (1500 x 50 cm), grey PVC base tanks continually generate eggs. These tanks are meticulously monitored and kept in the best sanitary conditions. They are filled with filtered seawater (salinity 35 g/l, filtered to 1 μm) and provided with Rhodomonas algae as their food source. The tanks are maintained at 16–18 °C, and modest bottom-supplied aeration is used.

The base tanks are changed once a week by introducing stage IV-V copepodites to maintain adult concentrations with a male-to-female ratio of 1:1 at fewer than 100 per litre. Every day, new saltwater is substituted for the 10 litres or so of culture water that contains eggs that is collected from the bottom of the tanks. An average of 95,000 eggs are produced each day using a 40 m screen to remove the eggs from the effluent water, which corresponds to a fecundity rate of 25 eggs per female each day. The adults are gathered on a 180 m sieve and transported to clean, disinfected tanks after the base cultures are emptied and cleaned two to three times a year.

In the growing tanks, where densities can approach 6000 per litre, the gathered eggs are moved. After 24 hours, nauplii begin to hatch; after 48 hours of incubation, the average hatching percentage is 50%. After ten days, a combination of Isochrysis and Rhodomonas is given at concentrations of 570 and 900 cells/ml, respectively. Initially, Isochrysis is given at a dosage of 1000 cells/ml. With a steady mortality rate of around 5% per day, the generation time—the time needed to attain 50% fertilised females is roughly 20 days. The adults are gathered using a 180 m sieve after 21 days and put to the base or harvest tanks.

When the fish hatchery starts up, the harvest tanks come into action. These 450-liter black tanks are kept in the same conditions as the other tanks indicated above and have the same capacity. Daily deliveries of 16X108 Rhodomonas cells from blood cultures are made to each tank. Compared to stock tanks, the harvest tanks need to be cleaned and emptied more frequently. The eggs are collected daily and moved to the hatching tanks, where aeration levels are maintained to maintain 80% oxygen saturation, making it easier to harvest nauplii or copepodites at a given stage. After receiving Isochrysis feedings, nauplii of the right size are collected using a 45 m screen, which also reduces adult copepod cannibalism.

## References

Ansari, S. & Khan, A. A. (2014). Diversity and distribution of zooplankton in a freshwater body of Aligarh region. *Indian Journal of Scientific Research and Technology*, 2(1): 73-82.

Dai, Y., Yang, Y., Wu, Z., Feng, Q., Xie, S., & Liu, Y. (2016). Spatiotemporal variation of planktonic and sediment bacterial assemblages in two plateau freshwater lakes at different trophic status. *Applied Microbiology and Biotechnology*, 100, 4161-4175.

Das, P., Mandal, S. C., Bhagabati, S. K., Akhtar, M. S. & Singh, S. K. (2012). Important live food organisms and their role in aquaculture. *Frontiers in aquaculture*, 5(4), 69-86.

Findlay, D. L., & Kling, H. J. (2001). Protocols for measuring biodiversity: phytoplankton in freshwater. *Winnipeg: Department of Fisheries and Oceans*, 19.

Gireesh, R., Varghese, M., & Thomas, V. J. (2015). Phytoplankton-collection, estimation, classification and diversity.

Hagiwara, A., &Hanazato, T. (1999). Techniques for the culture of Cladocera. In L. A. Greenberg, C. A. C. Pederson, & P. J. Guerinot (Eds.), Protocols in Aquatic Macroecology (pp. 223-230). American Society of Limnology and Oceanography.

Kamble, S. P., Patil, S. R. & Babare, M. R. (2013). Seasonal diversity of protozoans, rotiferans, cladocerans, and copepodans, from Krishna River Ghat near Miraj, Dist. Sangli, MS, India. *Galaxy: International Multidisciplinary Research Journal*, pp. 1-7.

Manickam, N., Bhavan, P. S., Santhanam, P., Bhuvaneswari, R., Muralisankar, T., Srinivasan, V., ... &Karthik, M. (2018). Impact of seasonal changes in zooplankton biodiversity in Ukkadam Lake, Coimbatore, Tamil Nadu, India, and potential future implications of climate change. *The Journal of Basic and Applied Zoology*, 79, 1-10.

Snell, T. W., & Janssen, C. R. (Eds.). (2011). Rotifera, Volume 4: The Progesterone. Guides to the Identification of the Microinvertebrates of the Continental Waters of the World. Kenobi Productions.

Stelzer, C. P., & Snell, T. W. (Eds.). (2003). Culture of Protostome Animals: Rotifers. CRC Press.

Yadav. P. (2015). Studies on biodiversity of zooplankton in Kass lake during rainy and winter seasons. *International Journal of Researches in Biosciences, Agriculture and Technology*, 3(2): 399- 402.

CHAPTER

# Reservoir Fisheries in India

8

## Introduction

Reservoirs, manufactured lakes designed to impound river water, serve various purposes such as irrigation, power generation, flood control, and meeting industrial water requirements. Numerous river valley projects have been undertaken since our independence in the quest for development, resulting in artificial impoundments. These expansive water bodies possess immense potential for biogenic production and, due to their substantial size, play a crucial role as a valuable resource for developing inland fisheries in our country. India boasts extensive freshwater resources, including rivers, reservoirs, lakes, and ponds. The tropical climate of Indian pools contributes to their high primary productivity, enabling them to yield a greater abundance of fish compared to the current national average of 29.4kg/ha/yr in reservoirs. Reservoir fisheries primarily operate as a combination of stocking and capture systems. The country has 975 reservoirs covering a total area of over 3.15 million hectares.

Reservoir classification varies among different authors, primarily based on the size of the reservoirs and their water spread area. Mohanty (1984) identified three types of reservoirs:

- **Minor reservoirs:** These have a water spread area of up to 40 hectares.
- **Medium reservoirs:** Encompass a water spread area of up to 400 hectares.
- **Major reservoirs:** These are characterized by a water spread area exceeding 400 hectares.

**Pathak** (1990) classified reservoir in three categories:

- **Large reservoirs:** These cover an area of 5000 hectares or more.
- **Medium reservoirs** have an impounded water spread area of 1000 to 5000 ha.
- **Small reservoirs:** These have a water spread area of less than 1000 ha.

**Agarwal** (1990) classified reservoir into four categories based on availability and management factors:

- **Large reservoirs:** have a water spread area of 1000 to 5000 hectares.
- **Medium reservoirs:** encompass a water spread area between 100 and 1000 hectares.
- **Minor reservoirs:** have a water spread area between 10 and 100 hectares.
- **Small reservoirs** have a water spread area below 10 hectares.

**Jhingran and Sugunan** (1990) classified reservoirs into the following groups:

- **Large reservoirs** have a water spread area exceeding 1000ha.
- **Medium reservoirs:** They have a water spread area ranging from 500 to 1000 hectares.
- **Small reservoirs** have a water spread area below 500 hectares.

In the Government of India's records, reservoirs are often categorised as small (1,000 hectares), medium (1,000 to 5,000 hectares), and big (>5,000 hectares) (Sarma, 1990; Srivastava *et al.*, 1985).

## Yield of Reservoir

The fish yield in reservoirs, which ranges from 0.05 kg/ha in Bihar to 35.5 kg/ha in Himachal Pradesh, is mediocre with a national average of 20 kg/ha. The average output of 50 kg/ha from India's small reservoirs is still rather low, ranging from 3.9 kg/ha in Bihar to 188 kg/ha in Andhra Pradesh. Other countries, such as China (743 kg/ha), Sri Lanka (300 kg/ha), Indonesia (177 kg/ha), Cuba (100 kg/ha), and the USSR (88 kg/ha), produce significantly more fish than Thailand (64.5 kg/ha).

**The main reservoir in India**

| Name of Dam | State | River |
|---|---|---|
| Rihand | Uttar Pradesh | Rehnd (Ganga) |
| Dhandraul | Uttar Pradesh | Bhakra |
| Sardasagar | Uttar Pradesh | ChukaSanda (Ganga) |
| Dhora | Uttar Pradesh | Dhora |
| Matatila | Uttar Pradesh | Ganga |
| GovidSagar | Punjab & Himachal Pradesh | Sutlej |
| Beas | Punjab & Himachal Pradseh | Beas |
| Hirakud | Odisha | Mahanadi |
| Rana Pratap Sagar | Rajasthan | Chambal |
| Maithon | Bihar | Barakar (Ganga) |
| Panchet | Bihar | Damodar (Ganga) |
| GhandiSagar | M.P. | Chambal (Ganga) |
| Mettur | T.N | Cauvery |
| Bhavanisagar | T.N | Bhavani |
| NagarjunaSagar | Andhra Pradesh | Krishna |
| NizamSagar | Andhra Pradesh | Mowgina |
| Tungabhadra | Karnataka | Tungabhadra (Krishna) |
| KrishnarajSagar | Karnataka | Cauvery |
| Neyyar | Kerala | Neyyar |
| Ukai | Gujarat | Tapti |

## Limnological Profile of Indian Reservoirs

India is home to numerous reservoirs that serve multiple purposes, such as irrigation, hydropower generation, and water supply.

### 1. Physical Characteristics

- **Size and Depth:** Indian reservoirs vary in size and depth. Some are large and deep, while others are smaller and shallow.
- **Surface Area:** The surface area of reservoirs can range from a few to several thousand hectares.
- **Water Volume:** Reservoirs in India can hold vast amounts of water, with volumes ranging from millions to billions of cubic meters.
- **Shoreline Development:** The shoreline development of reservoirs can vary. Some reservoirs have extensive shorelines with irregular shapes, while others may have simpler geometry.

## 2. Water Quality

- **Temperature:** Reservoirs in India generally exhibit a seasonal variation in water temperature. The surface temperature tends to be highest in summer and lowest in winter.
- **Turbidity:** Turbidity levels can vary based on sediment input from inflowing rivers, monsoon rains, and anthropogenic activities in the catchment area.
- **Dissolved Oxygen (DO):** DO levels are crucial for the survival of aquatic organisms. Factors such as temperature, algal blooms, and organic matter decomposition can affect DO concentrations in reservoirs.
- **pH Level:** The pH of Indian reservoirs can range from slightly acidic to slightly alkaline, depending on geology, land use, and inputs from surrounding areas.
- **Nutrient Levels:** Reservoirs often receive nutrients from inflowing rivers, runoff, and anthropogenic sources. High nutrient concentrations can lead to eutrophication and algal blooms.
- **Pollution:** Reservoirs in India face challenges related to pollution, including industrial waste, agricultural runoff, and domestic sewage.

## 3. Biological Characteristics

In most reservoirs, the rapid flushing rate brought on by the monsoon's substantial water discharge delays the establishment of macrophytic populations. Indian reservoirs are distinguished for their diverse plankton communities.

**Plankton:** Indian reservoirs harbour diverse planktonic communities, including phytoplankton (microscopic algae) and zooplankton (tiny animals). Blooms of harmful algae are sometimes observed. Indian reservoirs exhibit a diverse range of plankton communities. Plankton refers to the microscopic organisms that drift in water bodies, including phytoplankton (photosynthetic organisms) and zooplankton (animal-like organisms). The dominant plankton species in Indian reservoirs can vary depending on the geographical location, climatic conditions, water quality, and nutrient availability.

- **Cyanobacteria:** Cyanobacteria, also known as blue-green algae, often dominate Indian reservoirs. They can form harmful algal blooms under favourable conditions, leading to water quality issues. Cyanobacteria are commonly observed in reservoirs across India, including the Chilika Lake in Odisha, Sardar Sarovar Reservoir in Gujarat, and Pulicat Lake in Tamil Nadu (Sharma & Rai, 2017).

- **Diatoms:** Diatoms are phytoplankton characterized by their intricate silica cell walls. They are commonly found in Indian reservoirs and contribute significantly to primary production. Diatoms have been reported in several reservoirs, such as the Hirakud Reservoir in Odisha and Bhadra Reservoir in Karnataka (Venkataraman&Biju, 2018)
- **Rotifers:** Rotifers are tiny, multicellular zooplankton organisms found in Indian reservoirs. They are essential to the planktonic food web, consuming bacteria, algae, and other small organisms. Rotifers have been observed in various reservoirs, including the Nagarjuna Sagar Reservoir in Andhra Pradesh and the Tehri Reservoir in Uttarakhand (Raghavan& Ali, 2013).
- **Copepods:** Copepods are tiny crustaceans essential to the zooplankton community. They are crucial in transferring energy from phytoplankton to higher trophic levels in aquatic ecosystems. Copepods have been reported in Indian reservoirs such as the Nagarjuna Sagar Reservoir and the Kabini Reservoir in Karnataka (Khan & Reddy, 2016).
- **Cladocerans:** Cladocerans, commonly known as water fleas, are another group of zooplankton found in Indian reservoirs. They serve as a food source for fish and other organisms. Cladocerans have been documented in reservoirs like the Hirakud Reservoir and the Ukai Reservoir in Gujarat (Sarma, *et al.*, 2005)

**Macrophytes**: Submerged, floating, and emergent aquatic plants can be found in Indian reservoirs. They provide habitats, influence nutrient cycling, and contribute to biodiversity. Macrophytes are a diverse group of aquatic plants that play a crucial role in the ecology and functioning of reservoirs. In India, reservoirs are essential water bodies for irrigation, drinking water supply, hydroelectric power generation, and fisheries. The presence of macrophytes in these reservoirs is important for maintaining water quality, providing habitat and food for aquatic organisms, and stabilizing sediment. The diversity of macrophytes in Indian reservoirs can vary depending on geographical location, climate, water quality, and reservoir history.

- **Kabini Reservoir, Karnataka:** Kabini Reservoir is located in the southern state of Karnataka. A study conducted by Chandrashekara and Ramachandra (2010) documented the presence of various macrophyte species, including Hydrillaverticillata, Eichhornia crassipes, Vallisneria spiralis, and Ceratophyllum demersum, among others.

- **Sardar Sarovar Reservoir, Gujarat:** Sardar Sarovar Reservoir is a large reservoir situated on the Narmada River in the state of Gujarat. A study by Prajapati *et al.* (2014) assessed the macrophyte diversity in the reservoir and identified several species, including *Nelumbo nucifera, Eichhornia crassipes,* and *Ipomoea aquatica.* The research provides information on the richness and distribution of macrophytes in the pool and their ecological significance.
- **Ukai Reservoir, Gujarat:** Ukai Reservoir is located on the Tapi River in the state of Gujarat. A study by Patel and Pandya (2013) investigated the macrophyte flora of the reservoir and documented the presence of species such as *Eichhornia crassipes, Pistia stratiotes,* and *Typha angustifolia.* The research provides insights into the macrophyte composition and the potential impact of invasive species on the reservoir ecosystem.

**Fish and Aquatic Fauna:** Reservoirs are often stocked with fish species for commercial and recreational purposes. They also support many native aquatic fauna, including amphibians, reptiles, and invertebrates.

- **Bhakra Dam Reservoir, Himachal Pradesh:** Fish species: The reservoir is known for supporting various fish species, including Mahseer (*Tor putitora*), Catla (*Catla catla*), Rohu (*Labeo rohita*), and Indian Carp (*Cyprinus carpio*).
- **Nagarjuna Sagar Dam Reservoir, Andhra Pradesh/Telangana:** Fish species: The reservoir is home to several fish species, such as Indian Major Carp (*Catla catla*), Rohu (*Labeo rohita*), Mrigal Carp (*Cirrhinus mrigala*), and Silver Carp (*Hypophthalmichthys molitrix*).
- **Sardar Sarovar Dam Reservoir, Gujarat:** Fish species: The reservoir supports a diverse range of fish species, including Mahseer (*Tor putitora*), Indian Major Carp (*Catla catla*), Rohu (*Labeo rohita*), Tilapia (*Oreochromis mossambicus*), and Hilsa (*Tenualosa ilisha*).
- **Hirakud Dam Reservoir, Odisha:** Fish species: The reservoir is known for housing various fish species, such as Indian Major Carp (Catla catla), Rohu (*Labeo rohita*), Mrigal Carp (*Cirrhinus mrigala*), and Catfish (*Clarias batrachus*).

It's important to note that the specific limnological characteristics of individual reservoirs in India can vary significantly based on factors such as geographical location, climate, catchment characteristics, and human activities in the surrounding areas.

**The primary difference between a small reservoir and a large reservoir**

*(Jhingran, 1988)*

| Small reservoir | Large reservoir |
|---|---|
| Reservoirs with a single use are often used for irrigation. | Reservoirs with several uses, such as for large-scale irrigation, hydroelectric power generation, and small flood control. |
| Dams are not very complex or expensive. On small seasonal streams, structures made of earth, stone, and masonry work. | Dams are complex structures constructed of cement, concrete, or stone that are erected on enduring or lengthy seasonal rivers with exact technical competence. |
| physiologically more productive per area in the shallows. While few in seasonal reservoirs, aquatic plants are numerous in perpetual ones. | less biologically productive per unit area in the deep. Aquatic vegetation are typically absent. I had frequent draw downs. |
| The water regime may vary noticeably in the summer when it completely dries out. | Do not fully dry up. The water regime changes gradually. Maintain a dead storage (conservation-pool) level. |
| Sheltered areas are absent. | Sheltered areas by way of embayments, coves, etc., present. |
| The shoreline wasn't too erratic. sections of the littoral that gently slope. | The shoreline was more irregular. Littoral areas are mostly steep. |
| In these shallow, non-stratified reservoirs with little wave action, oxygen is mostly produced by photosynthesis. | Although photosynthesis provides a source of dissolved oxygen, it is only possible in a restricted area known as the euphotic zone, which is defined by the vertical range of light transmission. Another source of oxygen is strong wave action.. |
| The structure's design and size are determined by the magnitude of the runoff and are provided with a concrete or stone spillway. | Provided with more complex engineering devices |
| Major carps breed in the reservoirs. | The headwaters or other appropriate reservoir locations are where breeding is most frequently seen. |
| It may be manipulated experimentally to explore different ecological responses to environmental changes. | It is ineligible for experimental modification. |
| Chemical therapy and drainage can prevent the trophic depression phase. The reservoir may be drained as many times as it takes to complete a fish production cycle. | The trophic depression phase sets in. |

| Small reservoir | Large reservoir |
|---|---|
| One may relate the annual floods to the floodplains overflowing during the rainy season. Inundation of dry land causes the release of nutrients into the reservoir as it fills, resulting in a high production of fish food through the breakdown of organic materials, primarily of plant origin, leading to higher fish growth and survival. | Nutrient loss happens because they are trapped in silt at the bottom. Production of benthos will be decreased by rapid sedimentation. |
| There is no longer any brood stock due to overfishing or total exploitation of seasonal reservoirs. Rebuilding the fish population requires stocking. Thus, a direct connection between stocking rate and catch per unit of effort is established. | High-density fingerling stocking is necessary to balance stock numbers against natural mortality since there are noticeable yearly recruitment swings. Effective capturing techniques are needed. |

## Reservoir Fisheries Management

Reservoir fisheries management involves planning and implementing strategies to sustainably manage fish populations in reservoirs. It aims to maintain healthy fish stocks, promote biodiversity, and provide recreational fishing opportunities.

### 1. Stocking and Species Selection:

Reservoirs are often stocked with fish to establish or enhance populations. Stocking programs typically involve selecting appropriate fish species based on the reservoir's characteristics and management objectives. Commonly stocked species include largemouth bass, trout, catfish, and sunfish.

### 2. Habitat Management:

Creating and maintaining suitable habitats is crucial for the success of fish populations in reservoirs. This involves managing vegetation, improving water quality, and implementing artificial structures like fish attractors or brush piles.

### 3. Harvest Regulations:

Establishing appropriate harvest regulations helps ensure sustainable fishing and maintain fish populations at desired levels. These regulations may include size and bag limits, fishing seasons, and catch-and-release policies.

### 4. Monitoring and Research:

Regular monitoring of fish populations is essential for effective reservoir fisheries management. It involves assessing fish abundance, growth rates, population demographics, and health. Research studies can provide valuable insights into the reservoir's ecology and guide management decisions.

### 5. Stakeholder Engagement:

Engaging with anglers, local communities, and other stakeholders is crucial for successful reservoir fisheries management. Stakeholders' input can help shape management plans, improve compliance with regulations, and promote understanding of the importance of sustainable practices.

### 6. Invasive Species Management:

Controlling and preventing the spread of invasive species is vital to maintain the ecological balance of reservoirs. Strategies include monitoring and early detection, implementing prevention measures, and conducting eradication or control efforts when necessary.

## References

Aquatic Invasive Species - Reservoir Management (https://www.nwf.org/Educational-Resources/Wildlife-Guide/Threats-to-Wildlife/Invasive-Species)

Chandrashekara, U. M., & Ramachandra, T. V. (2010). Aquatic macrophytes of Kabini reservoir in Karnataka, India: composition, distribution patterns and ecological implications. Aquatic Ecosystem Health & Management, 13(3), 243-252.]

Colorado Parks and Wildlife - Reservoir Fishing Opportunities. (https://cpw.state.co.us/thingstodo/Pages/Fishing.aspx)

Himachal Pradesh Fisheries Department. (n.d.). Bhakra Dam. Retrieved from http://hpfisheries.gov.in/bhakra-dam/

https://annamalaiuniversity.ac.in/studport/download/marine/resources/Aquaculture_in_Reservoiur.pdf

https://www.researchgate.net/publication/280243998_Reservoir_Fisheries_of_India

Inland Fisheries Research and Surveys - Reservoir Fish Population Studies (https://tpwd.texas.gov/fishboat/fish/management/stocking/inland/reservoir.phtml)

Khan, M. I., & Reddy, Y. R. (2016). Zooplankton diversity in freshwater lakes of India: a review of recent advances and future prospects. *Biodiversity and Conservation*, 25(7), 1303-1338.)

Mohanty, R. P. (1984). Search decision rule application in reservoir management. *J. Inst. Eng.(India), Mech. Eng. Div.;(India), 64.*

Nagarjuna Sagar Project. (n.d.). Fish Species. Retrieved from http://nagarjunasagar.co.in/fish-species/

Nagarjuna Sagar Project. (n.d.). Fish Species. Retrieved from http://nagarjunasagar.co.in/fish-species/

Narmada Control Authority. (n.d.). Fish Biodiversity. Retrieved from https://nca.gov.in/biodiversity-fish/

North Carolina Wildlife Resources Commission - Reservoir Habitat Management] (https://www.ncwildlife.org/Conserving/Habitat-Enhancement/Freshwater-Habitat-Enhancement/Reservoir-Habitat-Enhancement

Odisha Tourism. (n.d.). Hirakud Dam. Retrieved from https://www.odishatourism.gov.in/content/tourism/en/place/202-place.html

Patel, K. R., & Pandya, H. G. (2013). Macrophytic diversity in Ukai Reservoir, Gujarat, India. Journal of Threatened Taxa, 5(5), 3938-3942.

Prajapati, S. K., Parmar, M. G., &Barot, N. H. (2014). Study on macrophyte diversity of Sardar Sarovar reservoir, Gujarat. *Indian Journal of Ecology,* 41(1), 96-99.

Raghavan, R., & Ali, A. (2013). Zooplankton diversity in floodplain lakes of Assam, northeast India. *Journal of Threatened Taxa*, 5(2), 3673-3687.)

Sarma, S. S. S., Nandini, S., & Gulati, R. D. (2005). Interactions between nutrient loading, zooplankton grazing and phytoplankton species richness are related to lake morphometry: evidence from a deep temperate lake. Freshwater Biology, 50(3), 600-610.)

Sharma & Rai, 2017). Cyanobacteria in freshwater ecosystems of India: a review. Environmental Monitoring and Assessment, 189(10), 508.

Texas Parks and Wildlife Department - Reservoir Stocking (https://tpwd.texas.gov/fishboat/fish/management/stocking/inland/reservoir.phtml)

Venkataraman, G., &Biju, A. (2018). Diatoms of Indian reservoirs. Springer, Singapore.)

CHAPTER 9

# Types of Lakes

## Introduction

A lake is a standing body of water located within a basin and isolated from the ocean (Forel, 1982). Additionally, he described a pond as a lake with shallow depths and a swamp as a pond with shallow depths containing rooted vegetation. A pond is a tranquil body of water with floating plants extending toward the center, similar to the littoral zone of a lake. Carpenter's theory (1928) suggests that the key distinction between a lake and a pond lies in their depth rather than their surface area (Larson & Schaetzl, 2001).

## Lake Morphology

The process of formation plays a significant role in determining the physical characteristics of a lake basin. The depth and contour of the lake bottom can be measured using methods such as a weighted line or, more efficiently, an echosounder. The physical structure of lakes is influenced by their shape, as well as how light and heat are distributed and how water circulates within them. To assess the presence of harmful compounds in a lake and compute the dynamics of nutrients, the hydraulic retention time, which is the time required for all of the lake's water to flow out its outlet, must be measured. The hydraulic retention time is mostly controlled by the lake's water supply and basin structure (Silva *et al.*, 2018).

## Lake Zonation

Lakes are categorized into distinct zones:

- The coastal zone runs from the shoreline, just beyond the reach of waves and spray, to a depth where there is just enough light to support the growth of rooted plants.

- The coastal area, which is typically well-mixed, extends from the lake surface to a point where the light level is only one percent of that at the surface, known as the photic (euphotic) zone.
- The aphotic zone lies below the littoral and photic zones at the lake's bottom, where light levels are too low for photosynthesis. Respiration occurs at all depths in the aphotic zone, leading to the consumption of oxygen. This dark and deep area is also referred to as the profundal zone.
- The compensation dept is the depth at which respiration and photosynthesis rates are balanced.
- The sublittoral zone, the deepest zone for plant growth, is situated between the profundal and littoral zones.
- The limnetic zone refers to the upper layer of surface water located offshore, away from the influence of the shoreline.

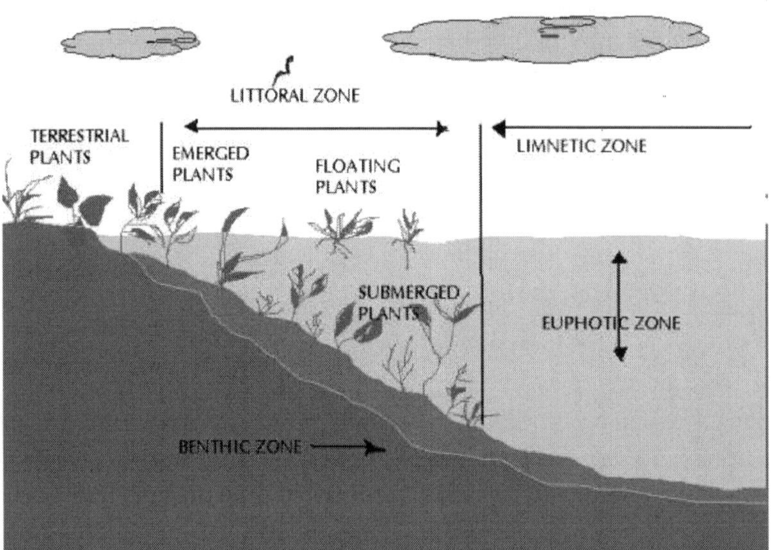


**Fig. Lake Zonation**

*(https://www.researchgate.net/publication)*

## Thermal Classification of Lakes

Lakes can be categorized based on variations in surface water temperature, as outlined by Hutchinson (1957):

- **Amictic:** Top and bottom waters remain unmixed, insulated by ice covers, and unaffected by weather or external factors.

- **Monomictic:** These lakes experience a single annual mixing of their waters, typical in most deep lakes on Earth.
- **Cold Monomictic:** These lakes never reach a depth of 4°C because they are ice-bound or covered in ice during the winter. They have inverse thermal stratification, with surface waters at 0°C and bottom waters at 4°C. As observed in Polar lakes, there is only one mixing event, which occurs at temperatures not exceeding 4°C in the spring and summer.
- **Warm monomictic:** Regardless of depth, the water temperature never drops below 4°C. These lakes undergo direct thermal stratification, with surface waters ranging from 10–20°C and bottom waters from 8–4°C. They experience a single annual mixing event, typically in the winter, as seen in most subtropical deep lakes.
- **Polymictic:** Continuous mixing occurs only at low temperatures.

## Thermal stratification

In tropical lakes, heat absorption at the surface creates a vertical temperature gradient where thermal resistance becomes too strong for prevailing winds to mix the entire water column effectively (Sly *et al.*, 1992). During the summer, three distinct layers emerge: the epilimnion (upper layer), the hypolimnion (lower layer), and the thermocline or metalimnion (middle layer).

### Epilimnion:

- This upper layer is the warmer portion of the water column.
- The temperature of this layer varies depending on atmospheric conditions, typically ranging from 27°C to 21°C.

### Hypolimnion:

- Referring to the water's base, this layer is characterized by cooler temperatures.
- The temperature in this layer typically ranges from 5°C to 7°C.
- The water column remains still within the hypolimnion.

### Thermocline:

- Situated in the middle as the metalimnion, the thermocline exhibits temperatures between those of the upper and lower layers.
- It's known for its gradient of temperature from the top to the bottom and is often referred to as a transition zone.

Deeper lakes display seasonal thermal phenomena that exert significant and far-reaching effects, directly and indirectly shaping the foundation on which the entire biological ecosystem depends, especially within the temperature range. Therefore, a comprehensive understanding of thermal stratification's key aspects is essential.

## Classification of lakes bases on their origin

### 1. Tectonic Lakes

Tectonic lakes originate from the Earth's crust undergoing lateral and vertical movements due to geological deformations, including faulting, folding, tilting, and warping. Rift lakes, exemplified by those in the Lake Baikal and Central African Rift regions, are among the world's largest lakes and are situated within rift valleys. Other notable tectonic lakes, such as the Caspian Sea, the Sea of Aral, and several lakes in the Ponto-Caspian region, result from basins being isolated from the ocean due to tectonic uplift of the seafloor above sea level.

**Fig. Tectonic Lake**

*(https://www.lakescientist.com/lake-origins)*

### 2. Volcanic Lakes (Crater Lakes)

Volcanic lakes are found in local depressions such as craters and maars, as well as larger basins produced by volcanic activity such as calderas. When precipitation accumulates faster than it can be drained through evaporation, groundwater discharge, or a combination of these causes, crater lakes occur. These lakes are also known as caldera lakes at times. Crater Lake, located within the caldera of Mount Mazama, collapsed around 4860 BCE as a result of a massive volcanic eruption. When lava flows or volcanic lahars stop rivers or streams, volcanic lakes form. As an example, Malheur Lake in Oregon was formed when a lava flow obstructed the Malheur River. Volcanic crater lakes are often almost round in shape.

**Fig. Glacial Lake**
*(https://www.theguardian.com/environment/2021/may/02/glacial-lakes)*

**Fig. Volcanic Lake**
*(https://en.wikipedia.org/wiki/Kelimutu)*

## 3. Glacial Lakes

Glacial lakes form as a direct result of glaciers and continental ice sheets. diverse glacial processes produce diverse sorts of basins, making it difficult to distinguish them from lakes generated by other sources. Morainic and outwash lakes, lakes in direct contact with ice, glacially carved rock basins and depressions, and glacial drift basins are the most common types of glacial lakes. Glacial lakes are found all over the world, with the majority of them in northern Europe and North America, where they were either shaped by or wholly formed by the most recent glacier that covered the area. Proglacial, subglacial, finger, and epishelf lakes are all examples of glacial lakes. Epishelf lakes have a high degree of stratification, with a freshwater layer formed by melting ice and snow and maintained in place by an ice shelf connected to the sea.

## 4. Basal Rock Dissolution Lakes

Basal rock dissolution lakes form when soluble rock, such as calcium carbonate, slowly dissolves in water. These lakes are observed in regions like California and the United States, where clear lakes exhibit this phenomenon.

## 5. Oxbow Lakes

Oxbow lakes are crescent-shaped bodies of water situated on valley floors adjacent to rivers or streams. Over time, these lakes form due to changes in the river's course resulting from erosion and sediment deposition. Water flows increasingly inside the river's loop, depositing silt, while the outside banks crumble, deepening the meander. When a river bends and separates from its main flow, oxbow lakes occur. Some of these lakes may become completely isolated, with only occasional water flow, particularly during floods. Notable examples include Dal and Wular lakes in Kashmir.

**Fig. Glacial Lake**

*(https://www.theguardian.com/environment/2021/may/02/glacial-lakes)*

**Fig. Volcanic Lake**

*(https://en.wikipedia.org/wiki/Kelimutu)*

## 6. Fluvial Lakes

Fluvial lakes are the result of river activity and form along the course of rivers. These lakes come into existence as slow-moving streams bend and meander, creating horseshoe bends. Ultimately, the river narrows at a certain point.

**Fig. Fluvial Lake**

*(https://link.springer.com/referenceworkentry/10.1007/978-1-4614-9213-9_463-2)*

## 7. Aeolian Lakes

Aeolian lakes, also known as wind-generated lakes, are mostly found in arid places, however some of them are remains of past arid climates. Depression basins carved by wind action in ancient desert settings, interdunal lakes located between well-oriented sand dunes, and lake basins obstructed by wind-blown sand are all examples of these lakes. Moses Lake in Washington is an example of a lake basin that has been dammed by wind-blown sand.

## 8. Shoreline Lakes

Shoreline lakes are often generated as a result of estuary obstruction or the uneven accumulation of beach ridges caused by longshore currents and other causes. These lakes include marine coastal lakes, which are frequently found within drowned estuaries, as well as lakes that are isolated from larger bodies of water by bars or formed by the junction of two spits.

## 9. Reservoir lakes (Anthropogenic lakes)

Reservoir lakes, also known as anthropogenic lakes, are man-made bodies of water formed by human activity such as dam construction across rivers and streams. They can also arise as a result of abandoned excavations being filled with groundwater, precipitation, or a combination of the two. Notably, the Upper Silesian region in southern Poland has about 4,000 man-made lakes, each with a unique origin, such as dry reservoirs built by dams and flooded mines.

## 10. Bog lakes

Bog lakes thrive in the northern temperate glaciated zone, which has high humidity and plentiful year-round precipitation. Because of variables such as low soil temperatures, low evaporation rates, and limited runoff, these lakes support strong plant growth. Bog lakes often have open water portions surrounded by natural boundaries and may have peat deposits at the edges or at the bottom.

**Fig. Reservoir lake**

*(https://stock.adobe.com/in/search?k=zervreilasee)*

**Fig. Shoreline lakes**

*(https://videohive.net/item/shoreline-lakes/22017405)*

## 11. Salt Lakes

Salt lakes are numerous worldwide, and their water composition resembles that of freshwater lakes. Climate change-induced dry conditions or geological changes in drainage basins can significantly reduce the annual inflow of water into these lakes. This reduction can lead to a substantial outflow decrease, causing the lake to transform into a sink or terminal lake. Ultimately, the lake may dry up as salts from incoming streams concentrate due to evaporation. Salt lakes are prevalent in arid regions, which includes the Great Salt Lake in Utah, Lake Walker, and the Pyramid in Nevada, and they can be located in a variety of dry climates around the world, including Australia, South America, East Africa, Antarctica, Russia, the northern Himalayas, South America, and North East America.

**Fig. Bog Lake**

*(https://guides.nynhp.org/bog-lakepond)*

**Fig. Salt lake**

*(https://www.britannica.com/place/Great-Salt-Lake)*

## 10. Bog lakes

Bogs flourish in the north temperate glaciated zone when precipitation is plentiful year-round and air humidity is high. As a result of the low soil temperature, decreased evaporation, and minimal runoff water, there is ample plant growth. A typical bog lake is an expanse of open water enclosed entirely or partially by genuine margins. Having peat deposits around the edges or in the bottom, typically with a fake bottom of extremely finely divided flocculent vegetative matter (Patrick *et al.*, 1981).

# References

Cael, B. B.; Heathcote, A. J.; Seekell, D. A. (2017). The volume and mean depth of Earth's lakes. *Geophysical Research Letters*. 44 (1): 209–218.

Cael, B. B.; Seekell, D. A. (July 8 2016). "The size-distribution of Earth's lakes". Scientific Reports. 6 (1): 29633.

Cohen, A. S. (2003). Paleolimnology: The History and Evolution of Lake Systems. *New York: Oxford University Press*. ISBN 978-0-19-513353-0.

Currey, Donald R. (1990). Quaternary palaeolakes in the evolution of semidesert basins, with particular emphasis on Lake Bonneville and the Great Basin, U.S.A. Palaeogeography, Palaeoclimatology, *Palaeoecology*. 76 (3–4): 189–214.

Downing, J.A.; Prairie, Y. T.; Cole, J.J.; Duarte, C.M.; Tranvik, L.J.; Striegl, R.G.; McDowell, W.H.; Kortelainen, P.; Caraco, N.F.; Melack, J.M. (2006). "The global abundance and size distribution of lakes, ponds, and impoundments". *Limnology and Oceanography*. 51 (5): 2388–2397.

Forel, F.A., 1901. Handbuch der Seenkunde. Allgemeine Limnologie. J. von Engelhorn, Stuttgart, Germany.

Håkanson, L & Jansson, M. (1983). Principles of Lake Sedimentology. Berlin: Springer-verlag. (Vol. 316).

Håkanson, Lars (2012). Lakes on Earth, Different Types. In: Bengtsson, Lars; Herschy, Reginald W.; Fairbridge, Rhodes W. (eds.). Encyclopedia of Lakes and Reservoirs. Dordrecht: Springer. pp. 471–472.

Hutchinson, G. E. (1957). A Treatise on Limnology. Vol.1, Geography, Physics, and Chemistry. New York: Wiley.

Koc, J.K., Kobus, S. and Glińska-Lewczuk, K., 2009: A significance of oxbows to the wooded river valleys ecosystems (in Polish). *Journal of Water Land Develeopement*, 13a: 115–131.

Larson, G., &Schaetzl, R. (2001). Origin and evolution of the Great Lakes. *Journal of Great Lakes Research*, 27(4), 518-546.

Last, W.M. and Smol, J.P. (2001). Tracking environmental change using lake sediments. Volume 1: basin analysis, coring, and chronological techniques. *Springer Science & Business Media.*

Leach, J. H., & Herron, R. C. (1992). A review of Lake Habitat classification. The development of an aquatic habitat classification system for lakes, 27-58.

Löffler, H. (2003). The origin of lake basins. The lakes handbook: Limnology and limnetic ecology, 8-60.

Matsubara, Yo; Howard, Alan D.; Drummond, Sarah A. (April 1 2011). "Hydrology of early Mars: Lake basins". *Journal of Geophysical Research*. 116 (E04001): E04001.

McDonald, Cory P.; Rover, Jennifer A.; Stets, Edward G.; Striegl, Robert G. (2012). "The regional abundance and size distribution of lakes and reservoirs in the United States and implications for estimates of global lake extent". *Limnology and Oceanography*. 57 (2): 597–606.

Patrick, R., Binetti, V. P. &Halterman, S. G. (1981). "Acid lakes from natural and anthropogenic causes". *Science*. 211 (4481): 446–8.

Rouwet, D. (2015). Volcanic Lakes, Advances in Volcanology, Springer-Verlag Berlin Heidelberg.

Rzętała, Mariusz; Jagus, Andrzej (May 2011). "New lake district in Europe: Origin and hydrochemical characteristics". *Water and Environment Journal.* 26 (1): 108–117

Seekell, David A.; Pace, Michael L. (2011). "Does the Pareto distribution adequately describe the size-distribution of lakes?". *Limnology and Oceanography.* 56 (1): 350–356.

Sen Nag, Oishimaya (2020). Is the Caspian sea a sea or a lake?. worldatlas.com. WorldAtlas.

Silva, M. S., Guimaraes, J. T., Souza Filho, P. W., Nascimento Junior, W., Sahoo, P. K., Costa, F. R. & Costa, M. F. (2018). Morphology and morphometry of upland lakes over lateritic crust, Serra dos Carajás, southeastern Amazon region. *Anais da Academia Brasileira de Ciências,* 90, 1309-1325.

Sly, P. G., Busch, W. D. N., & Busch, W. D. N. (1992). A system for aquatic habitat classification of lakes. The development of an aquatic habitat classification system for lakes, 15-26.

Wang, Zhen-Ting, Tian-Yuan, Chen, Liu, Si-Wen, Lai &Zhong-Ping (March 2016). "Aeolian origin of interdune lakes in the Badain Jaran Desert, China". *Arabian Journal of Geosciences.* 9 (3): 190

Wetzel, Robert (2001). Limnology: Lake and river ecosystems. San *Diego: Academic Press.* ISBN 9780127447605.

Witham, Fred, Llewellin &Edward W. (2006). "Stability of lava lakes". *Journal of Volcanology and Geothermal Research.* 158 (3–4): 321–332.

CHAPTER 10

# Prominent Lakes of India Subcontinent

## Indian lakes

A lake is a naturally occurring, sizable body of water that is situated in a basin that is encircled by dry ground. A lake often has a slower flow than the stream(s), which serve as its inflow or outflow. Although lakes are entirely on land and are not connected to the ocean, they contribute to the Earth's water cycle by acting as sizable standing water reservoirs, much like the much bigger seas. Most lakes are freshwater, making up nearly all of the surface freshwater on Earth, but others are salt lakes with salinities much higher than those of the ocean. There are no official definitions or scientific criteria differentiating the two, but ponds and lakes are water-filled basins on land, albeit ponds are often more extensive and more profound.

Most lakes receive both their inflow and outflow through creeks and rivers, although some lakes are endorheic and have no flow. In contrast, volcanic lakes receive their influx only from precipitation runoffs and lack any inflow streams. Lagoons are small tidal ponds at the shore blocked off by sandbars. Lagoons are different from lakes. Natural lakes are typically found in hilly regions (such as alpine lakes), dormant volcano craters, rift zones, and areas still experiencing glacial. A river channel that has broadened over a basin created by eroded floodplains and wetlands can be found in depressed landforms or along the courses of mature rivers.

The numerous lakes in some regions of the planet were created by the erratic drainage patterns left over from the last ice age. Over an extended period, all lakes are transient because they gradually fill with sediment or overflow their containment basin. Reservoirs are manufactured lakes that are controlled artificially. They are typically built for industrial or agricultural usage, hydroelectric power generation, home drinking water supply, ecological or recreational purposes, or other human activities.

Numerous lakes may be found in nearly every region of the country, which significantly enhances their natural appeal. Unquestionably, Rajasthan and Himachal Pradesh, the northern and western states of India, respectively, have more lakes than any other region.

**Fig. Famous lakes in India**

*(Source: Authors)*

## 1. Dal Lake, Jammu Kashmir

Dal Lake is a lake in Srinagar, the capital of Jammu and Kashmir. Kashmir's second-largest urban lake is known as Srinagar's Jewel and is referred to as the "Jewel in the Crown of Kashmir" or "Srinagar's Jewel" since it is so important to tourism and pleasure in the region. The lake is crucial for commercial fishing and water plant harvesting operations (Khan *et al.*, 2012). A promenade surrounded by gardens, parks, hotels, and houseboats from the Mughal era runs the length of the lake's shoreline, roughly 15.5 kilometres (9.6 miles). The lake freezes throughout the winter since the temperature occasionally drops below 11 °C (12 °F). A natural wetland that includes the lake, which has a surface size of 18 square kilometers (6.9 square miles), and its floating gardens, has 21.1 square kilometers (8.1 square miles).

In the Kashmiri region, there are floating gardens known as "Rad" that blossom with lotus flowers during the months of July and August. These gardens are separated by causeways into four distinct basins: Gagribal, Lokut Dal, Bod Dal, and Nigeen (although Nigeen is sometimes considered a separate lake). Within Lokut Dal and Bod Dal, there are islands known as Rupa Lank (or Char Chinari) and Sona Lank, respectively. Efforts are underway to address the significant issue of eutrophication in Dal Lake and its surrounding Mughal gardens, with the Indian government investing a substantial $275 million (11 billion rupees) to restore the lake to its natural state.

## 2. Brahma Sarovar, Haryana

In Thanesar, in the Indian state of Haryana, there is a water tank known as Brahma Sarovar that is revered by the Dharmic religions. Bathing is essential for maintaining internal and exterior purity in dharmic religions (Deswal and Dogra, 2023).

## 3. Bhimtal Lake, Uttarakhand

This lake, which can be found 22 miles from Nainital, bears the name of Bhima, the second Pandava in the illustrious Mahabharata. It ranks as one of the major lakes in Nainital and the second-largest lake in Kumaoun. At a height of 1,375 meters (4,511 feet), the lake is situated in the Uttarakhand town of Bhimtal in the Nainital district. The lake drains a catchment area of 17.12 square kilometers (6.61 square miles). Boating, fishing, and angling are all excellent activities that may be enjoyed on the Lake (Panwar and Malik 2016).

## 4. Roopkund Lake, Uttarakhand

Roopkund is a glacial lake nestled in the high-altitude region of Uttarakhand, India, often referred to locally as either Mystery Lake or Skeleton Lake. This stunning lake is situated within the Trishul massif, a prominent mountain range in the Himalayas. Roopkund Lake is perched at a lofty altitude of approximately 5,020 meters (equivalent to 16,470 feet), and it is encircled by majestic snow-clad mountains and glaciers. It is renowned as a popular trekking destination, attracting adventurers and nature enthusiasts. The size of Roopkund Lake can vary significantly, but it typically remains relatively small, seldom exceeding a diameter of 40 meters, equivalent to an area of 1000 to 1500 square meters. During the winter months, the lake typically freezes over, adding to its mystique and allure.

## 5. Upper Lake - Madhya Pradesh

Upper lake (*Bhojtal*) is located in the western portion of the city of Bhopal, the capital of Madhya Pradesh. It is bordered to the south by the Van Vihar National Park, the east and north by populated areas, and the west by agricultural fields. It has a 31 km$^2$ area and a 361 km$^2$ catchment or watershed that it drains (Upadhyay *et al.*, 2012). It is a significant drinking water supply for the city's inhabitants, providing over 30 million imperial gallons (140,000 m$^3$) of water daily to almost 40% of the population, now a Ramsar site.

## 6. Jaisamand Lake, Rajasthan

Dhebar Lake, commonly known as Jaisamand Lake, holds the distinction of being India's second-largest artificial freshwater lake and the world's oldest historical man-made lake. It is situated in Rajasthan State's Udaipur District in western India. It was created at Namla Thikana in the 17th century when Rana Jai Singh of Udaipur constructed a marble dam across the Gomati River. At its full expanse, Jaisamand Lake covers an area spanning 87 square kilometers (equivalent to 34 square miles). It is roughly 45.0 kilometres (28.0 miles) from Udaipur's district administrative centre. It was biggest artificial lake in the world when it was initially constructed. To reach the Jaisamand Wildlife Sanctuary that surrounds Dhebar Lake, travelers can take the state highway leading to Banswara from Udaipur. Salumbar, a sub-district headquarters on State Highway No. 32, is about 17.0 km (10.6 miles) away. The Dhebar Lake shoreline is protected by the Jaisamand Wildlife Sanctuary, which covers around 162.0 square kilometres (16,200 ha), most of which is teak forest. Within the lake, there are three islands varying in size, spanning from 10 to 40 acres (equivalent to 40,000 to 162,000 square meters). Notably, the Dhebar Lake Marble Dam, measuring an impressive 300.0 meters (equivalent to 984.3 feet) in length, is recognized as a "Heritage Monument of India." Additionally, the dam also houses the Hawa Mahal Palace, which once served as the winter capital of the Maharanas of Mewar.

## 7. Pushkar Lake- Rajasthan

Pushkar Lake, located in Rajasthan, is an artificial lake that was formed by building a dam across the headwaters of the Luni River in the 12th century. It holds great religious significance for Hindus in India. Pushkar Lake is also famous for the annual Pushkar Camel Fair, one of the largest livestock fairs in the world. It usually takes place in the month of November and attracts tourists and traders from all over the world. It's a colorful and vibrant event that includes camel races, cultural performances, and religious rituals (Mathur *et al.*, 2007).

## 8. Pulicut Lake, Andhra Pradesh

The area of Pulicat Lake, which is 759 square kilometers (293 square miles), makes it the second largest brackish water lagoon in India after Chilka Lake. The lagoon is the most significant portion is in Andhra Pradesh's Tirupati district. The lagoon is one of three essential wetlands that, from October to December, draw rain clouds from the northeast monsoon. The lagoon's boundary only extends to around 96% of Andhra Pradesh and 3% of Tamil Nadu. The lagoon has sand ridges covering its western and eastern halves, parallel to the coastline. The lagoon's size varies depending on the tide; at high tide, it covers 450 square kilometers (170 square miles), while at low tide, it covers 250 square kilometers (97 square miles). It measures roughly 60 kilometers (37 miles) in length and has a breadth that ranges from 0.2 kilometres (0.12 miles) to 17.5 kilometres (10.9 miles). Tropical monsoons are the primary weather factor along the lagoon shore, from 15 °C (59 °F) to 45 °C (113 °F), the ambient temperature ranges. Sriharikota, a sizable barrier island with a spindle-like appearance, divides the lagoon from the Bay of Bengal.

## 9. Hebbal Lake- Karnataka

Hebbal Lake is located in the northern part of Bangalore, at the junction of Bellary Road and the outer ring road, where National Highway 7 begins. This lake has a historical significance as it was one of the three lakes commissioned by KempeGowda in 1537. Its formation followed the typical method of creating lakes or "tanks" in the Bangalore region, involving the construction of bunds to block natural valley systems. In 2000, a survey indicated that the lake spanned an area of 75 hectares, with plans in place to expand it to 143 hectares. The lake's catchment area was found to be approximately 3,750 acres. Over the years, the lake's size had diminished, measuring 77.95 hectares in 1974 and 57.75 hectares in 1998. The annual catchment volume was estimated at 15.2 million cubic meters, distributed through the Northeast Monsoon (3.04 million cubic meters), Southwest Monsoon (10.12 million cubic meters), and the dry season (3.28 million cubic meters). Through desilting efforts, the lake's storage capacity was increased from the anticipated 2.38 million cubic meters in 2000 to 4.07 million cubic meters. Unfortunately, the lake's chemical and biological composition has undergone changes due to the inflow of sewage, impacting its ecosystem (Jalilzadeh *et al.*, 2007).

## 10. Powai Lake, Maharashtra

An artificial lake, Powai Lake is located in the Powai Valley, a northern Mumbai suburb, below Vihar Lake on the Mithi River. Powai is the name of the lake in the city suburb. Over time, there has been a noticeable growth in population near the lake. The lake's water spread area was around 2.1 kilometers (520 acres) at the time of construction, and its depth ranged from roughly 3 meters (9.8 feet) near the edge to 12 meters (39 feet) at its deepest point. Powai Lake's water quality has deteriorated over numerous phases. The lake water that supplied drinking water to Mumbai has been deemed unsafe. The lake is still famous among travelers (Sharma and Uchchariya, 2020).

## 11. Loktak Lakes, Manipur

Loktak Lake, located near Moirang in the Indian state of Manipur, stands as the world's only floating lake. It holds the distinction of being northeastern India's largest freshwater lake and was originally one of the three lakes commissioned by KempeGowda in 1537. The lake's unique feature is its floating phumdis, consisting of a diverse mixture of plant life, soil, and organic matter in various stages of decomposition. The name "Loktak" originates from the etymological roots denoting "stream" and "the end." This historic lake plays a vital role in Manipur's economy, serving purposes such as irrigation, drinking water supply, and electricity generation. Local fishermen residing in the vicinity, including on the phumdis, known as "phumshongs," derive their livelihoods from the lake. Loktak Lake was accorded international significance as a wetland under the Ramsar Convention on March 23, 1990, a recognition based on its ecological condition and the rich biodiversity it harbors (Roy and Majumder 2019).

## 12. Chilka Lakes, Odisha

Chilka Lake, a brackish water lagoon situated in the Indian state of Odisha, encompasses regions within the districts of Puri, Khurda, and Ganjam along the eastern coast, where the Daya River converges with the Bay of Bengal. The lagoon is the second largest in the world and the largest coastal lagoon in India. It is the most significant migratory bird wintering area on the Indian subcontinent. The catchment area of Chilka Lake spans 3,560 square kilometers, and it is surrounded by a wealth of endangered plant and animal species. There are abundant fisheries resources in the lake's environment. It provides food for more than 150,000 fishermen who live in 132 settlements on the coast and islands. The Chilika Lagoon's brackish water is home to microalgae, marine seaweeds, grasses, fish, and crabs (Annandale and Kemp 1915).

## 13. Hussain Sagar, Andhra Pradesh

Hussain Sagar, or Tank Bund, is a heart-shaped lake in Hyderabad, Telangana, constructed by Ibrahim Quli Qutb Shah in 1563. It covers 5.7 square kilometers (2.2 squre miles) and is nourished by the river Musi. In the center of the lake, on Gibraltar Rock, is a huge monolithic monument of the Gautama Buddha, built in 1992. Additionally, it divides Secunderabad from Hyderabad's suburbs in the city core. The lake's deepest point is 32 feet (9.8 m) deep (Javed *et al.*, 2008).

## 14. Vembanad Lake, Kerala

The longest Lake in India is the Vembanad Lake, 96.5 kilometers long. Vembanad Backwater's area decreased from 230 km$^2$ in 1968 to 179.25 km$^2$ in 1983. Alappuzha, Kottayam, and Ernakulam districts are all about the lake. It is at sea level, with a thin barrier island separating it from the Laccadive Sea. Vembanad Lake is interconnected with neighboring coastal lakes through a network of canals. Notable islands like Pathiramanal, Perumbalam, and Pallippuram are ensconced within its waters (Sruthy and Ramasamy 2017). At its widest point, the lake measures 14 kilometers in breadth. Vembanad Lake is a crucial component of the Vembanad-Kol wetland system and is fed by ten rivers, including the Achenkovil, Manimala, Meenachil, Muvattupuzha, Pamba, and Periyar. These rivers collectively drain an area of 15,770 square kilometers, accounting for 40% of Kerala's total land area and supplying nearly 30% of the state's annual surface runoff, equivalent to 21,900 million cubic meters.

## 15. Osman Sagar Lake

Osman Sagar Lake, affectionately known as "Gandipet," is an artificial reservoir created by the construction of a dam on the Isa River, a tributary of the Musi River. As documented by Balijepalli in 2009, this man-made lake plays a pivotal role as the primary freshwater source for the twin cities of Secunderabad and Hyderabad. Moreover, Osman Sagar Lake also supports thriving fisheries, contributing to the local economy and providing a source of livelihood for fishermen in the region.

## 16. Kodaikanal Lake - Tamil Nadu

Kodaikanal Lake, also known as Kodai Lake, is a man-made lake nestled in the city of Kodaikanal, located in Tamil Nadu's Dindigul district, India. This lake stands out as one of Kodaikanal's most prominent natural features and a favored destination for tourists. It features a well-established boat club, boathouse, and public boat service, adding aesthetic value to tourism. The lake hosts annual events such as the Boat Pageant and Flower Shows, which draw tourists, particularly during the summer months (Balamurugan *et al.*, 2014).

## References

Annandale, N. and Kemp, S. (1915). Fauna of the Chilka Lake. *Memoirs of the Indian Museum*, 5(1), 1-20.

Balamurugan, P., Vasudevan, S., Selvaganapathi, R. and Nishikanth, C. V. (2014). Spatial distribution of grain size characteristics and its role in interpreting the sedimentary depositional environment, Kodaikanal Lake, Tamil Nadu, India. *Journal Earth Science and Climate Change*, 5(217), 2.

Balijepalli, V. P. (2009). Analysis of drinking water supply system encompassing the catchment, the reservoir and the treatment facility (A case study of Osman sagar drinking water supply system, Hyderabad, India).

Cross, S. L., Baker, P. A., Seltzer, G. O., Fritz, S. C. and Dunbar, R. B. (2001). Late Quaternary climate and hydrology of tropical South America inferred from an isotopic and chemical model of Lake Titicaca, Bolivia and Peru. *Quaternary Research*, 56(1), 1-9.

Deswal, S.and Dogra, S. K. (2023). Water Quality Assessment of Brahma Sarovar: A Sacred Lake at Kurukshetra (India). *Ecological Questions*, 34(3), 1-15.

Dolan, D. M. and Chapra, S. C. (2012). Great Lakes total phosphorus revisited: 1. Loading analysis and update (1994–2008). *Journal of Great Lakes Research*, 38(4), 730-740.

Jalilzadeh, A. K., Yamakanamardi, S. M. and Altaff, K. (2007). Abundance of zooplankton in three contrasting lakes of Mysore city, Karnataka state, India. In: *Proceedings of Taal 2007: The 12th World Lake Conference (Vol.* 464, p. 469).

Javed, S. M., Reddy, N. C., Srinivasulu, C., Kodarkar, M. S., Reddy, C. S. and Tampal, F. (2008). Restoration and enhancement of Hussain Sagar lake biodiversity. *In Proceedings of International Lake Workshop, JICA-HUDA, Hyderabad, India.*

Khan, J. A., Gavali, R. S. and Shouche, Y. S. (2012). Exploring the present status of hydrochemistry and sediment chemistry of Dal Lake, Kashmir and the effect of anthropogenic, disturbances on it. *Indian Journal Innovative Develeopment*, 1(7), 554-571.

Mathur, P., Agarwal, S. and Nag, M. (2007). Assessment of physicochemical characteristics and suggested restoration measures for Pushkar Lake, Ajmer Rajasthan (India). In: *Proceedings of Taal 2017: The 12th World Lake Conference* (Vol. 1518, p. 1529).

Panwar, S. and Malik, D. S. (2016). Zooplankton diversity, species richness and their distribution pattern in Bhimtal Lake of Kumaun region,(Uttarakhand). *Hydrology Current Research*, 7(1), 219.

Roy, R. and Majumder, M. (2019). Assessment of water quality trends in Loktak Lake, Manipur, India. *Environmental earth sciences*, 78, 1-12.

Sharma, D. K. and Uchchariya, R. (2020). Water quality assessment of Pagara reservoir, Joura, Morena district, Madhya Pradesh. *Indian Journal of Scientific Research*, 10(2), 93-102.

Shivaji, S., Bhadra, B., Rao, R. S. and Pradhan, S. (2008). *Rhodotorula himalayensis sp.* nov., a novel psychrophilic yeast isolated from Roopkund Lake of the Himalayan mountain ranges, India. *Extremophiles*, 12, 375-381.

Sruthy, S. and Ramasamy, E. V. (2017). Microplastic pollution in Vembanad Lake, Kerala, India: the first report of microplastics in lake and estuarine sediments in India. *Environmental pollution*, 222, 315-322.

Swann, G. E., Mackay, A. W., Vologina, E., Jones, M. D., Panizzo, V. N., Leng, M. J. and Sturm, M. (2018). Lake Baikal isotope records of Holocene central Asian precipitation. *Quaternary Science Reviews*, 189, 210-222.

Upadhyay, R., Pandey, A. K., Upadhyay, S. K., Bassin, J. K. and Misra, S. M. (2012). Limnochemistry and nutrient dynamics in Upper Lake, Bhopal, India. *Environmental monitoring and assessment*, 184, 7065-7077.

CHAPTER

# Riverine Fisheries in India

# 11

## Introduction

The river holds immense significance as both the lifeline and bloodline of India. The intricate network of main streams and their tributaries, collectively known as the "Drainage System" or "River System," forms the foundation of this vital resource. India is fortunate to be blessed with many rivers and tributaries, which contribute to the agricultural sector and are a crucial component of the country's inland transport system. A river is a natural watercourse that flows freely, typically containing freshwater, and finds its way toward an ocean, sea, lake, or another river. These rivers exhibit tremendous diversity in size, ranging from small streams that can be crossed on foot to massive water bodies stretching hundreds or even thousands of miles. Rivers play an indispensable role in the earth's ecosystems and are essential for the sustenance of life. The significance of rivers extends beyond their role as water sources. They provide drinking water, facilitate irrigation for agriculture, and supply water for industrial purposes. Furthermore, rivers serve as habitats for various species of fish, plants, and wildlife, ensuring their survival and fostering biodiversity. In the grand scheme of the water cycle, rivers are pivotal in maintaining the earth's climate by transporting heat and moisture from one region to another.

Human civilizations have thrived alongside rivers throughout history due to their numerous benefits. Rivers provide fertile land for agriculture, enabling the growth of crops and sustenance for communities. They also serve as vital transportation routes, facilitating trade and commerce by offering convenient channels for moving goods and people. In summary, rivers hold a position of paramount importance in India and the world. They are essential for sustaining life and contribute significantly to various aspects of human existence, ranging from agriculture to transportation. Recognizing the value of rivers and ensuring their conservation is crucial for the well-being and prosperity of current and future generations.

## Rivers in India

India has a vast river network that is an integral part of its culture, history, and economy. India's river systems comprise 14 major (catchments >20,000 $km^2$), 44 mediums (catchments 2,000 to 20,000 $km^2$), and 153 small rivers (catchments area <2,000 $km^2$) with a combined length of 45000 Km. This includes 25 basins and 101 sub-basins, and Only 4 of the 14 major rivers are completely perennial; the rest dry up in summer. The river is classified into different types below:

- **Classification of Rivers:** Rivers are categorized in several ways depending on the criteria used. Some common types of rivers:
- **Classification based on source:** Rivers can be classified as spring-fed or snow-fed, depending on their source. Spring-fed rivers originate from underground springs, while snow-fed rivers begin as melting snowpacks.
- **Classification based on flow:** Rivers can be classified as perennial or intermittent, depending on their flow. Perennial rivers flow all year round, while intermittent rivers only flow during certain times of the year.
- **Classification based on size:** Rivers can be classified as small, medium, or large, depending on their size.

### Classification based on the water flowing

Indian rivers based on the flowing classified into two categories: Himalayan flowing rivers and peninsular rivers.

## Major River Systems in India

### Ganga River System

The Ganges, stretching over 2525 kilometres and encompassing a drainage area of 9,51,600 square kilometers, which accounts for 25% of India's total surface, is the country's most extensive river system. Originating from the 'Alakananda' and 'Bhagirathi'sheadstreams at 'Garhwal' and the Gangotri glacier at Gomukh, respectively, the river eventually converges at 'Devaprayag' to form the Ganges. This mighty river sustains diverse aquatic life, including the endangered Ganges River dolphin, and supports a thriving and varied fish population. With a combined length of approximately 8047 kilometres, the Ganges river system holds immense significance as India's most important and ranks among the largest worldwide. The Ganges River flows through five states along its primary route: Uttarakhand, Uttar Pradesh, Bihar, Jharkhand, and West Bengal. According to the National Ganga River Basin Authority (NGRBA) in 2011, the Ganges basin encompasses not only these five states but also extends to include Delhi, Haryana, Himachal Pradesh, Madhya Pradesh, Chhattisgarh, and Rajasthan.

| Tributaries of Ganga Right Bank | Distance (km) | Tributaries of Ganga Left Bank | Distance (km) |
|---|---|---|---|
| Yamuna river | 1376 | Ramganga river | 569 |
| Chambal river | 1080 | Ghaghra river | 1080 |
| Banas river | | Meghna | |
| Sind river | | Gandak river | 300 |
| Betwa river | | BurhiGandak | |
| Ken river | 360 | Gomti river | 940 |
| Son river | 784 | Kosi river | 492 |
| | | Tamsa | |
| | | Mahananda | |
| | | Brahmaputra | 4000 |

The Ganga harbour has 265 fish species, of which 181 are recorded for the freshwater sector (Handbook of Fisheries and Aquaculture, 2019). According to a recent study conducted by the Central Inland Fisheries Research Institute, dated 2017, it has been asserted that the Ganga River's primary course, stretching from Tehri in Uttarakhand to Fraserganj in West Bengal, potentially houses a total of 158 distinct fish species.

*(Source:https://www.google.com/search)*

Furthermore, the middle and lower regions of the Ganga River, extending into the deltaic zone, offer favorable environments for a total of 14 testudines, encompassing 13 distinct turtle species and one tortoise species (ZSI, 1991). Among these, there are nine types of turtles characterized by their hard shells that can be encountered within the Ganga River ecosystem. These include *Batagur kachuga, B. dhongoka, B. baska, Hardella thurjii, Pangshura tecta, P. smithii, Geoclemys hamiltonii, Melanochelys trijuga,* and *P. tentoria,* with the latter exhibiting two subspecies, namely *P. t. tentoria* and *P. Nilssonia gangetica*. Additionally, *N. hurum, Chitra indica,* and *Lissemys punctata,* the latter boasting two subspecies, *L.P.P. punctata* and *L.P. andersoni,* represent the remaining four soft-shell turtle species inhabiting this region of the Ganga River.

- **Threatened fish species:** In the Ganga River, numerous endangered fish species find a habitat, including the Ganges river dolphin (*Platanista gangetica*), the Garial (*Gavialis gangeticus*), and the Golden mahseer (*Tor putitora*). These creatures face the threat of extinction primarily because of habitat degradation and excessive fishing activities.
- **Invasive species:** Numerous invasive fish species, including Common carp (*Cyprinus carpio*) and Nile tilapia (*Oreochromis niloticus*), have also invaded the Ganga River, harming the native fish population.

## Yamuna River

The Yamuna River, situated in North India, serves as a significant tributary of the Ganges, coursing its way through the region. It originates in the Yamunotri Glacier in the Himalayas and merges with the Ganges at Prayagraj. The Yamuna is approximately 1,376 km long and is a sacred river in Hinduism. It supports a rich and diverse fish fauna, but unfortunately, due to anthropogenic activities, its water quality has deteriorated, affecting fish diversity. The Yamuna River basin has documented a total of 105 different fish species, distributed across 53 genera and spanning 19 different families (Mir *et al.*, 2018).

- **Threatened fish species:** The Yamuna River helps protect endangered fish like Ganges river dolphins (*Platanista gangetica*), Gharials (*Gavialis gangeticus*), and Indian Softshell Turtles (*Aspideretes gangeticus*) from habitat loss and overfishing.
- **Invasive species:** The Yamuna River has also been invaded by several exotic fish species, such as the Common carp (*Cyprinus carpio*) and the Tilapia (*Oreochromis niloticus*), which have had negative impacts on the native fish fauna.

## Gomti River

The Gomti River, a river in India, serves as a significant tributary of the Ganges River, coursing through the northern Indian state of Uttar Pradesh. This river is renowned for its abundant and diverse fish population, providing essential sustenance and serving as a hub for various fishing endeavors, both for subsistence and commercial purposes. During the study period, 44 fish species belonging to 34 genera, 18 families and five orders were recorded. Cyprinids were the dominant group (39%), followed by Bagrid catfishes (16%) and Channids (8%). Exotic species, namely, *Cyprinus carpio* and *Oreochromis niloticus,* were recorded—endangered fish species *Chagunius chagunio,* and vulnerable species *Eutropiichthys vacha* (ICAR-CIFRI 2017).

## Brahmaputra River System

The Brahmaputra is a transboundary river that flows through China, India, and Bangladesh. It flows through the North-eastern states of India and is approximately 2,900 km long. The Brahmaputra, also called the Yarlung Tsangpo in Tibet, is a vital water source for irrigation and hydropower. It's known for its fertile plains and supports diverse fish life.

The river's north bank boasts tributaries like Depi, Dekari, Simen, Gainadi, Jiadahal, Subansiri, Ranganoi, Dikrong, Buroi, Burgang, Jiabharali, Belsiri, Pasnoi, Jiadhansiri, Puthimari, Pagladia, Beki, Manas, Aie, Champamati, Gaurang, Tipkai, Gadadhar, and Sankosh. On the south bank, you'll find Noadihing, Dibru, Burhidihing, Disang, Diknow, Dhansiri, Kopili, Kalong, Kulsi, Singra, Dudhnoi, and Krishnoi as notable tributaries. Combined with its 47 major tributaries, the river stretches for approximately 4000 kilometres, covering a catchment area of 5,80,000 square kilometres—the average water discharge measures around 5,10,450 cubic meters per second.

As per information from the National Bureau of Fish Genetic Resource (NBFGR), the Brahmaputra River basin has documented a grand total of 182 fish species, encompassing 64 different families (Dey, 2018). In contrast, the Handbook of Fisheries and Aquaculture reports a broader diversity, indicating the presence of 221 fish species representing 103 genera and 36 families in the same region, 2019).

- **Threatened fish species:** The Brahmaputra River also supports many threatened fish species, such as the Ganges river dolphin (*Platanista gangetica*), the gharial (*Gavialis gangeticus*), and the Chinese paddlefish (*Psephurus gladius*), which are endangered due to habitat loss and overfishing.

- **Invasive species:** Numerous foreign fish species, such the Common carp (*Cyprinus carpio*) and Tilapia (*Oreochromis niloticus*), have also entered the Brahmaputra River, harming the local fish population.

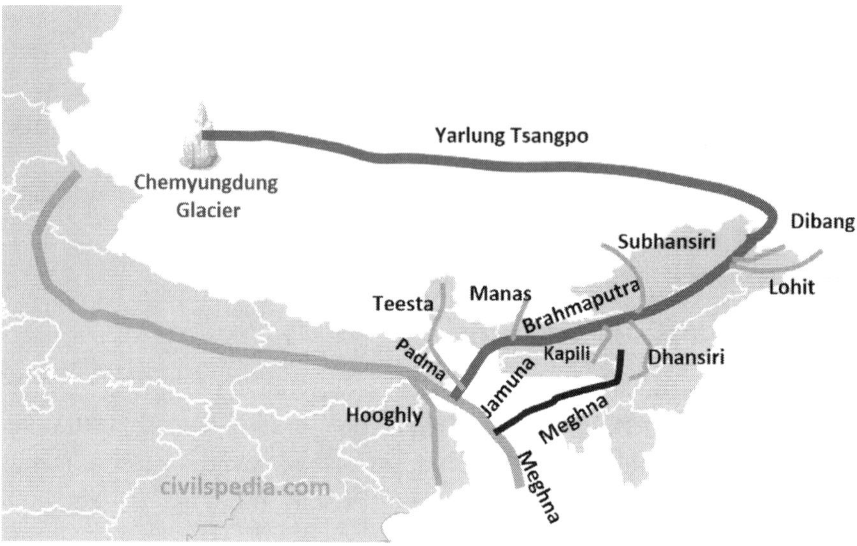


*(Source:https://www.google.com/search)*

## Indus River System

The Indus River, one of the most extensive waterways in Asia, has its origins in Tibet and meanders through the territories of India, Pakistan, and China. In the Indian subcontinent, specifically in the region of Jammu and Kashmir, it stretches approximately 1,114 kilometers in length. What sets the Indus River system apart is its westerly position on the subcontinent. It boasts several significant tributaries, including the Jhelum, Chenab, Ravi, Beas, and Satluj. The Indus River is responsible for nourishing a vast expanse of glaciers and mountain slopes, encompassing the Karakoram, Ladakh, Zanskar, and Himalayan Ranges. Its journey commences at the Bokhar Chu glacier on the northern slopes of Mt. Kailash (standing at 6714 meters) in Tibet. The river's course starts at the majestic Kailash Mountain, where it flows through Tibet's Singi Khamban valley, also known as the Lion's Mouth valley, in a northwesterly direction. As it enters Ladakh, the Indus River follows a remarkably straight path, situated between the Ladakh and Zanskar ranges. For the initial 480 miles, it cuts through flat terrain at elevations exceeding 3200 meters. Below Leh, the Zaskar River joins the Indus on its left bank, and two other left-bank tributaries, Suru and Dras, converge with the Indus near Kargil. Continuing on its northwest course, the Indus River encounters the Shyok-Nubra tributaries, originating from the Siachen Glacier in the Karakoram Range. These tributaries merge with the Indus as it continues its journey.

The Shigar River joins the Indus not far from Skardu, close to the Shyok River. Gilgit, another important tributary, connects to the Indus from the west. The Indus passes through a significant synclinal gorge as it crosses the middle Himalayan range. It goes southwest, passes Nanga Parbat, and enters Pakistan. There are at least 150 fish species in 23 families found in the Indus River System. The following fish species are some of the more prevalent ones in the Indus River System: Catla (*Catla catla*), Rohu (*Labeo rohita*), Silver carp (*Hypophthalmichthys molitrix*), Grass carp (*Ctenopharyngodon idella*), Common carp (*Cyprinus carpio*), Mahseer (*Tor spp.*), Featherback fish (*Notopterus spp*). The main types of fish discovered in the Indus River encompass *Schizopygopsis stoliczkei, Labeo gonius, Rita Buchana, Sisor rhabdophorus, Exostoma stolicizkae, Trichogaster*, and various *Nemacheilus* species.

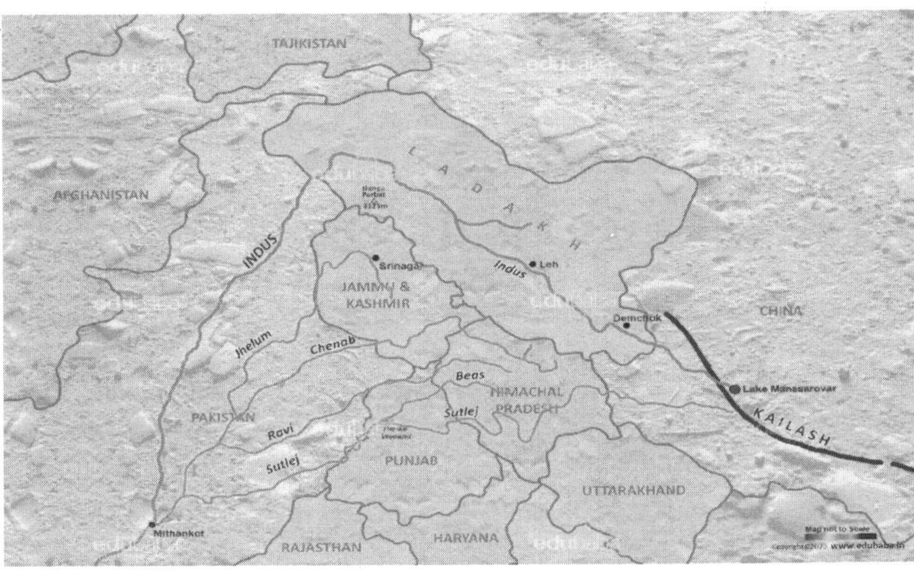


(*Source:https://www.google.com/search*)

## Chenab River Basin (Tributaries of Indus river)

The river Chenab, often known as Chandra Bhaga, is formed at the point where two streams, Chandra and Bhaga, meet in Tandi. Chandra and Bhaga originate from the Lahul and Spiti valleys in Himachal Pradesh, which are situated in the Himalayas. Chandra flows from the southwest side of the Barelacha Pass, while Bhaga flows from the northwest side. Chandra embarks on its journey by heading southeast for approximately 88 kilometers, tracing the foothills of the mid-Himalayas, before ultimately converging with Bhaga near Tandi. Even though the Bhaga River is merely 80 kilometers in length, its path leading up to the confluence is characterized by a steep slope with an average descent rate of roughly 24 meters per kilometer. Once these two streams merge, the resulting combined river is referred to as Chandra Bhaga, also known as Chenab. This united watercourse continues to flow for 48 kilometers in a northwesterly direction until it receives its first significant tributary, the Miyar Nallah, on the right bank.

After travelling another 96 kilometres, mostly northward, through Himachal Pradesh, it enters the Paddar neighbourhood of the Kishtwar district in the state of Jammu and Kashmir. It then travels through the Pangi valley. At Gulabgarh in the Padder Valley, it connects to the Bhutnallah. The river travels 61 km northwest from there until its major tributary, the Marusudar, enters it near Bhandarkot on the right bank. The river flows downstream in a southerly direction for about 32 km until it reaches Thathri, where it then turns west. The NiruNallah enters the Chenab River in this section, roughly 15 km downstream of Thathri, on its left bank. The river proceeds to flow an additional 41 kilometers in a northwesterly course until it encounters the Bischeleri tributary on the right bank. Subsequently, it continues its journey for approximately 64 kilometers to the west. During this stretch, numerous small streams converge with the Chenab River, including Cheneni, Talsuen, Yabu, and Ans on the right bank, as well as Katu Nallah, Mandial, and Painthal Khad on the left. At the point downstream of the Ans River's confluence, the Chenab changes its direction and travels southward for approximately 55 kilometers until it reaches Akhnoor. From there, it proceeds towards the Sialkot region of Pakistan. The river covers a distance of about 410 miles between the confluence of Chandra and Bhaga and Akhnoor. Notably, *Dipthychus maculates* and *Glyptothorax kashmirensis* are common species inhabiting the Chenab River.

## Ravi River (Tributary of Indus river)

The Ravi River has its beginnings in the Himalayan state of Himachal Pradesh's Chamba region. It travels 725 miles before draining 14,442 square km of India's catchment region. The PirPanjal and Dhauladhar mountains form a triangle on either side of its westward movement. The river runs constantly all year long and travels in a northwesterly direction. The Ravi River, the tiniest among the quintet of Punjab Rivers, commences its journey from the glacial expanses located on the southern incline of the Mid Himalayas, standing tall at an impressive elevation of 14,000 feet (equivalent to 4,300 meters). In its nascent stages, this river boasts brisk currents, with its riverbed strewn with an assortment of stones. Roughly 40 miles (about 64 kilometers) downstream from its origin, the Ravi River merges with two substantial tributaries known as the Budhil and Nai or Dhona. Within the waters of this river, you can encounter an array of notable fish species, including *Amblypharyngodon mola, Barilius bendelisis, Carassius carassius, Labeo catla, Cirrhina mrigala, C. reba, Cyprinus carpio, Labeo spp, Tor tor, Mastacembelus armatus, Channa spp., Trichogaster fasciatus, Clarias batrachus, Heteropneustes fossilis, Mystus spp., Ompok bimaculatus, O. padba, Rita rita,* and *Wallago attu.*

## Satluj River (Tributary of Indus basin):

The Satluj River, one of the major five rivers historically associated with the Punjab region, serves as a vital water source for several Indian states, namely Himachal Pradesh, Punjab, Haryana, Delhi, and Rajasthan. This river, which is a tributary of the Indus, originates from the Mansarover Lake situated in Tibet at an elevation of 4572 meters above sea level. Over its course, spanning 320 kilometers from the border, the Satluj River flows through various regions including Shipkilla, Pooh, Powari, Tapri, Karcham, Nathpa, Rampur, Sunni, and Dehar before reaching the Gobindsagar reservoir (Bhakra) located in Bilaspur, Himachal Pradesh.Notably, the Sutlej River has contributed to the biodiversity of the regions it traverses, housing a total of 70 species of aquatic life. Among these species, the following are particularly prominent: *Wallago attu, Clarius batrachus, Hetetropneustes fossilis, Mystus spp., Rita rita, Eutropiichthys vacha, Notopterus notopeterus, N. chitala, Puntius spp., Colisa fasciatus, Esomus danricus,* and *Chela laubuca.*

# West- coast Flowing Rivers

## Narmada River

The Narmada River flows through the regions of Gujarat, Maharashtra, and Madhya Pradesh, originating from the Maikal Hills and spanning a distance of approximately 1,312 kilometers. One of the main rivers in central India, the Narmada River is home to a large and diversified fish population. In the Narmada River watershed, 141 fish species from 63 genera and 23 families have been identified (Shukla & Patel, 2018).

- **Threatened fish species:** The Narmada River plays a crucial role in the conservation of several vulnerable fish species, including the Ganges river dolphin (*Platanista gangetica*), the Indian Freshwater Pearl Mussel (*Lamellidens corrianus*), and the Smooth-coated otter (*Lutrogale perspicillata*). These remarkable creatures face the threat of extinction primarily because of habitat degradation and excessive fishing activities.

- **Invasive species:** The Narmada River has also been invaded by several exotic fish species such as the Common carp (*Cyprinus carpio*) and the Tilapia (*Oreochromis niloticus*), which have had negative impacts on the native fish fauna.

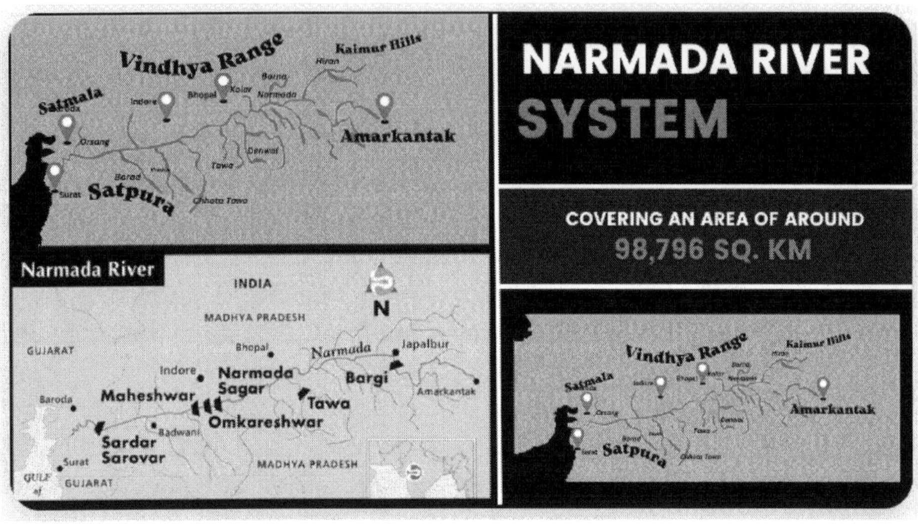


(Source:https://www.google.com/search)

## Tapti River

The prominent waterway known as the Tapti River holds the distinction of being the principal river in the western regions of India. This river traverses the states of Maharashtra, Gujarat, and Madhya Pradesh as it flows through its course. The river is known for its rich fish diversity, supporting subsistence and commercial fishing activities. Eighty-eight fish from 43 genera and 14 families have been recorded in the Tapti River basin (Rana, & Singh, 2018).

- **Threatened fish species:** The Tapti River plays a crucial role in supporting several at-risk aquatic species, including the Ganges River Dolphin (*Platanista gangetica*), the Indian Freshwater Pearl Mussel (*Lamellidens corrianus*), and the Smooth-coated Otter (*Lutrogale perspicillata*). These species are endangered owing to habitat degradation and exploitation.
- **Invasive species:** The Tapti River has also been invaded by several exotic fish species such as the Common carp (*Cyprinus carpio*) and the Tilapia (*Oreochromis niloticus*), which have had negative impacts on the native fish fauna.

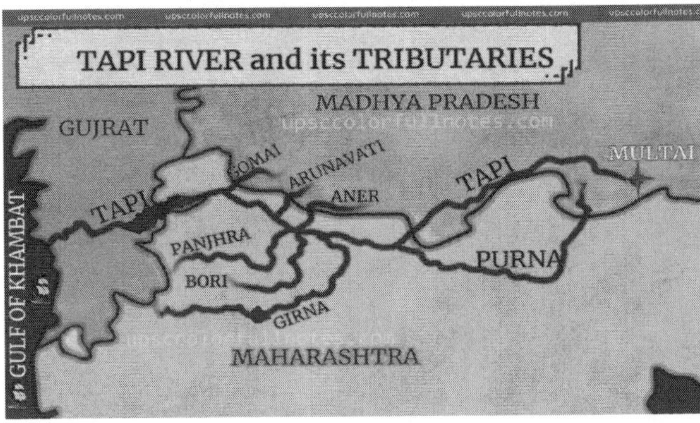


*(Source:https://www.google.com/search)*

# East Flowing River

## Mahanadi River

The Mahanadi is a river that flows through the states of Chhattisgarh and Odisha in Eastern India. It is approximately 858 km long and is an important water source for irrigation, hydropower, and industrial use. The Mahanadi is also home to the Hirakud Dam, one of the longest dams in the world, which was built to control flooding and generate hydroelectric power. The river is known for its rich fish diversity, supporting subsistence and commercial fishing activities. In the Mahanadi River basin, a comprehensive survey has identified a grand total of 174 different fish species. These fish are categorized into 78 distinct genera and are distributed across 24 diverse families (Mohapatra *et al.* 2018) (Handbook of Fisheries and Aquaculture, 2019) There are 253 species of fish in all, which are divided into 73 families.

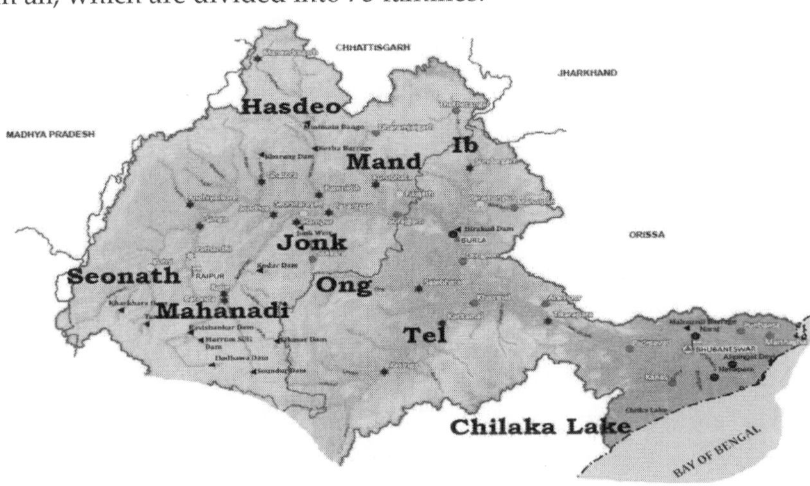


*(Source:https://www.google.com/search)*

- **Threatened fish species**: The Mahanadi River also supports many threatened fish species such as the Ganges river dolphin (*Platanista gangetica*), the Indian Freshwater Pearl Mussel (*Lamelliden scorrianus*), and the Olive Ridley Sea Turtle (*Lepidochelys olivacea*), which are endangered due to habitat loss and overfishing.
- **Invasive species**: Numerous foreign fish species, like the Common Carp (*Cyprinus carpio*) and the Tilapia (*Oreochromis niloticus*), have also invaded the Mahanadi River, harming the local fish population.

## Godavari River

The Godavari River, which ranks as the second-longest river in India, meanders through the Indian states of Maharashtra, Telangana, Andhra Pradesh, and Odisha. It originates in the Western Ghats and is approximately 1,465 km long. Two reservoirs (Gangapur and Nathsagar) are situated on the main stream of river Godavari in Maharastra. The Godavari River basin is renowned for its abundant variety of fish, which play a crucial role in sustaining both subsistence and commercial fishing ventures. In this region, a remarkable 173 different fish species, encompassing 76 genera and 22 families, have been documented (Rana & Singh, 2018).

*(Source:https://www.google.com/search)*

- **Threatened fish species**: The Ganges River Dolphin (*Platanista gangetica*), the Indian Freshwater Pearl Mussel (*Lamellidens corrianus*), and the Smooth-coated Otter (*Lutrogale perspicillata*), all of which are endangered owing to habitat degradation and exploitation, are just a few of the vulnerable fish species that the Godavari River sustains.

- **Invasive species**: Other foreign fish species that have invaded the Godavari River include the Common Carp (*Cyprinus carpio*) and the Tilapia (*Oreochromis niloticus*), which have had a detrimental effect on the local fish population.

## Cauvery River

The Cauvery River, one of the largest rivers in the southern regions of India, meanders its way through the states of Kerala, Tamil Nadu, and Karnataka. Originating from the lofty Brahmagiri highlands nestled within Karnataka's Western Ghats, this mighty river stretches across a vast expanse of 850 kilometers, playing a crucial role as a primary source of water for drinking, electricity generation, and agriculture in the region. Beyond its utilitarian significance, the Cauvery River is renowned for its abundant and diverse fish population, sustaining both subsistence and commercial fishing endeavors. According to data provided by the National Bureau of Fish Genetic Resources (NBFGR), the Cauvery River basin hosts an impressive array of 97 distinct fish species. These species are further classified into 47 different genera and are distributed across 17 distinct families. In total, 80 species of fish can be found within this river's waters, representing 23 different families (Handbook of Fisheries and Aquaculture, 2019).

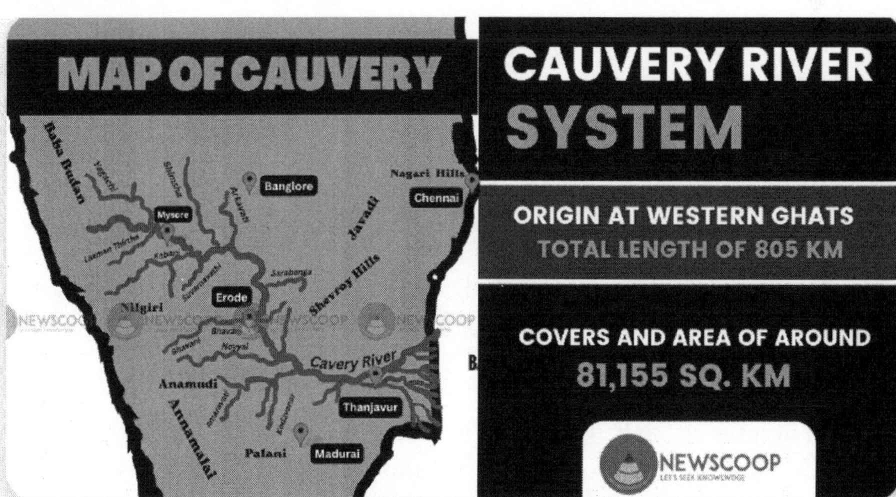


*(Source:https://www.google.com/search)*

- **Threatened fish species**: The Cauvery River also supports many threatened fish species, such as the Indian Freshwater Pearl Mussel (*Lamellidens corrianus*), the Vater's Catfish (*Glyptothorax vaterifloris*), and the Cauvery Mystus (*Mystus canarensis*), which are endangered due to habitat loss and overfishing.
- **Invasive species**: The Cauvery River has also been invaded by several exotic fish species, such as the African Catfish (*Clarias gariepinus*) and the Mozambique Tilapia (*Oreochromis mossambicus*), which have had negative impacts on the native fish fauna.

## Krishna River

One of the largest rivers in southern India, the Krishna River passes through the states of Maharashtra, Karnataka, and Andhra Pradesh. It has a total length of 1280 km, is fed by the Mahabaleshwar Hills in the Western Ghats of Maharashtra, and is a vital source of water for irrigation, hydropower, and drinking water supply. The river is known for its rich fish diversity, supporting subsistence and commercial fishing activities. According to the NBFGR, 102 fish species belonging to 48 genera and 17 families have been recorded in the Krishna River basin.

*(Source:https://www.google.com/search)*

- **Threatened fish species:** The Krishna River also supports many threatened fish species, such as the Ganges river dolphin (*Platanista gangetica*), the Indian Freshwater Pearl Mussel (*Lamellidens corrianus*), and the Hump-backed Mahseer (*Tor remadevii*), which are endangered due to habitat loss and overfishing.

- **Invasive species:** The Krishna River has also been invaded by several exotic fish species, such as the African Catfish (*Clarias gariepinus*) and the Common Carp (*Cyprinus carpio*), negatively impacting native fish fauna.

## Fishing Gear used in Rivers

The river is an essential source of fish in India, and several fishing gear are used to catch fish there. Here are some of the commonly used fishing gear in the Ganges River:

- **Cast nets:** Cast nets are circular nets that are thrown over a school of fish—the net spreads when it hits the water, trapping the fish inside.
- **Gill nets:** Gill nets are walls of netting that hang vertically in the water. Fish swim into the net and become entangled in the mesh.
- **Seine nets:** Seine nets are long nets pulled through the water by a boat. The netting is held open by floats and weights, trapping fish inside.
- **Traps:** Traps are usually made of woven bamboo or other materials and are used to catch fish that swim near the riverbank. Fish swim into the trap and cannot find their way out.
- **Hooks and lines:** Hooks and lines are used to catch larger fish, such as catfish and carp. The hooks are baited with pieces of fish or other bait, and the lines are cast into the water and left to sit until a fish bites.

## Major threats to Fishery

The major threats to fisheries resources in rivers can vary depending on the specific location and circumstances. Common hazards that are often associated with river fisheries include:

- **Overfishing:** Overfishing occurs when the fish population is harvested at a rate that exceeds its ability to reproduce and replenish itself. This can deplete fish stocks and disrupt the balance of the ecosystem (Cooke *et al.*, 2021).
- **Habitat Destruction:** The destruction and degradation of river habitats can significantly impact fish populations. Activities such as dam construction, channelization, deforestation, and pollution can alter the physical and chemical characteristics of the water, making it unsuitable for fish survival and reproduction (Moyle & Mount., 2007).
- **Pollution:** Pollution from various sources, including industrial discharges, agricultural runoff, and improper waste disposal, can

contaminate river ecosystems and negatively affect fish populations. Pollutants such as heavy metals, pesticides, and chemical nutrients can accumulate in fish tissues, leading to physiological and reproductive problems (Vajargah *et al.*, 2020).

- **Invasive Species:** Introducing non-native species into river ecosystems can disrupt the natural balance and threaten native fish populations. Invasive species often outcompete native species for resources, prey on them, or introduce diseases, leading to declines in fish populations (Ricciardi *et al.*, 2017).
- **Climate Change:** Climate change significantly threatens river fisheries. Rising temperatures, altered precipitation patterns, and changes in river flows can impact the availability of suitable habitats for fish species. Additionally, climate change can influence the timing of fish migrations, disrupt food webs, and increase the prevalence of diseases (Pörtner *et al.*, 2019).

## The Proliferation of Exotic Species in the Ganga River

The proliferation of exotic species in the Ganga River is a significant concern for its native fish population and the entire river ecosystem. Exotic species are non-native species introduced into an ecosystem intentionally or unintentionally. They can significantly impact native species and the ecosystem, leading to ecological imbalances and disruptions.

The Ganga River has witnessed the proliferation of several non-native species, among them being the African catfish, scientifically known as *Clarias gariepinus*, alongside the silver carp, scientifically designated as *Hypophthalmichthys molitrix*, and the common carp, scientifically referred to as *Cyprinus carpio*. These species have been introduced into the river for aquaculture and fishing purposes. Still, they have established themselves in the river and compete with native fish species for food and habitat.

Exotic species can also spread diseases and parasites to native species, leading to their decline and loss of biodiversity in the river ecosystem. Furthermore, the proliferation of exotic species can lead to losing traditional fishing practices and cultural values associated with native fish species.

To address the proliferation of exotic species in the Ganga River, there is a need for effective management and control measures. This can include strict regulations on introducing non-native species into the river, monitoring of aquaculture and fish stocking practices, and awareness campaigns to educate the public on the impact of exotic species on the river ecosystem. Additionally, restoration and conservation of native fish populations and their habitats can help to maintain the ecological balance of the river ecosystem.

Some of the exotic species that have been reported in the Ganga River include:

- African catfish (*Clarias gariepinus*)
- Silver carp (*Hypophthalmichthys molitrix*)
- Common carp (*Cyprinus carpio*)
- Nile tilapia (*Oreochromis niloticus*)
- Grass carp (*Ctenopharyngodon idella*)

These exotic species have been introduced into the Ganga River for aquaculture and other purposes, but they can negatively impact the native fish species and the river ecosystem.

## References

Cooke, S.J., Hogan, Z.S., Butcher, P.A. *et al.* (2021). Overfishing is likely to worsen, unless fisheries subsidies are reformed. *Nat Food* 2, 270–273.

Dey, S. (2018). Fish biodiversity and conservation in Brahmaputra River Basin, Northeast India. In: Freshwater Fishes of India (pp. 55-74). Springer, Singapore.

Ganga River System. National Institute of Hydrology. http://nihroorkee.gov.in/ganga-river-system/.

Gupta, S. K. (2018). Fish Diversity and Fishery Potential in the Ganga River Basin, India. In: Freshwater Fishes of India (pp. 121-135). Springer, Singapore.

ICAR-CIFRI (2017). Annual Report 2016-17. ICAR-Central Inland Fisheries Research Institute, Barrackpore, 232 pp.

Jaiswar, A.K., Das, P., Srivastava, A., 2020. Exotic fishes in the Ganga River system, India: A review. *Journal of Applied Ichthyology* 36, 979–987.

Javed, M.N., Ahmed, N., Hussain, M. *et al.* Fish diversity and conservation status of the Indus River Basin, Pakistan. *Journal of Applied Ichthyology*, 36, 406–420 (2020).

Mir, J. I., Chandra, S., &Bhagat, Y. (2018). Fish diversity and fisheries of the Yamuna River Basin, India. In Freshwater Fishes of India (pp. 101-120). Springer, Singapore.

Mohapatra, A., Das, M. K., & Chakraborty, S. K. (2018). Fish diversity and conservation in Mahanadi River Basin, Eastern India. In: Freshwater Fishes of India (pp. 153-174). Springer, Singapore.

Moyle, P.B., Mount, J.F. Riverine Fishes of California. University of California Press (2007).

NBFGR (2012). National Bureau of Fish Genetic Resources, Annual Report 2011-12. Lucknow, India.

NGRBA (National Ganga River Basin Authority) (2011). Environmental and Social Management Framework (ESMF). Environmental and Social Analysis, Vol. 1. New Delhi: Ministry of Environment, Forest & Climate Change, Government of India.

Pandey, A.K. and Mishra, A.K. (2011). Fish fauna of Gomti River with reference to conservation of threatened species, Uttar Pradesh, India. *Journal of Threatened Taxa*, 3(5), 1756-1761.

Pörtner, H.O., Roberts, D.C., Masson-Delmotte, V. *et al*. IPCC, 2019: Summary for Policymakers. In: IPCC special report on the ocean and cryosphere in a changing climate (2019). www.ipcc.ch/srocc/chapter/summary-for-policymakers/

Rana, K., & Singh, R. (2018). Fish diversity and conservation in Godavari River Basin, Eastern India. In: Freshwater Fishes of India (pp. 45-66). Springer, Singapore.

Rana, K., & Singh, R. (2018). Fish diversity and conservation in Tapti River Basin, Western India. In: Freshwater Fishes of India (pp. 25-44). Springer, Singapore.

Ricciardi, A., Blackburn, T.M., Carlton, J.T. *et al*. 2017). Invasion science: a horizon scan of emerging challenges and opportunities. *Trends of Ecology and Evolution* 32, 464–474.

Shukla, S. P., & Patel, H. (2018). Fish diversity and conservation in Narmada River Basin, Central India. In: Freshwater Fishes of India (pp. 75-100). Springer, Singapore.

Status of Fisheries in India - 2020. Department of Fisheries, Ministry of Fisheries, Animal Husbandry and Dairying, Government of India. https://vikaspedia.in/agriculture/fisheries/fisheries-industry-in-india/status-of-fisheries-in-india-2020.

Vajargah, M.F., Abdul Aziz, H., Ibrahim, Z. (2020) Effects of water pollution on fish diversity, abundance, and health: a review. *Environment Science Pollution Resourse*, 27, 10562–10579.

ZSI (1991). Faunal Resources of Ganga, Part I. Calcutta: Zoological Survey of India.

CHAPTER

# 12

# Traditional Fishing Crafts and Gears Used in India

## Introduction

Fishing is the skill of collecting fish and other aquatic animals. It is regarded as an ancient vocation with a lengthy history when humans primarily gathered sustenance from nature through water body fishing or wild animal hunting. In aquatic environments, fish are one of the principal exploitable resources that offer a cheap supply of protein. In addition to providing food, a significant portion of the rural community also receives nutritional security and a means of subsistence thanks to fishing (Sreekrishna and Shenoy, 2001). Artisanal fishermen's traditional fishing methods significantly influence Indian fisheries. To provide sustainability to the sector in the face of new problems in the capture fisheries, it is urgently necessary to have a better grasp of harvesting systems in addition to traditional fishing techniques. Traditional fishing is easier to implement, more energy-efficient, and environmentally benign. Additionally, compared to the mechanised sector, traditional fishing provides plenty of opportunity for the equitable distribution of ecological benifits (Hameed and Boopendranath, 2000).

## Fishing Crafts

The floating, mobile platform from which fishermen operate their equipment is called a fishing craft. The type of inland fishing vessels varies from location to location and is influenced mainly by the hydrological and topographical characteristics of the area. The fishermen frequently fished in various open waters using floating fishing vessels. A dugout canoe, a plank-built boat, a rubber tube platform, a bamboo raft, a thermocouple raft, a boat covered on a hardwood frame or bamboo frame, a large ship for group fishing, and a timber frame iron/tin/aluminium sheet boad are among the fishing vessels that have so far been documented to be used. The wooden camoes utilised in Madhya Pradesh are multi-purpose and used for fishing and transporting fish harvests (Kokate *et al.*, 2016).

## Dugout canoe

A little wooden canoe that has been dug out from a single piece of wood is known locally as a "Donga or Dongi." Sometimes, smaller canoes are constructed from palm tree trunks. Only one or two fishermen can fish simultaneously in a boat with an overall length of 3–4 m and a width of 0.5- 0.6m. This craft may set and carry fish traps, gill nets, cast nets, and lines in shallow water. Most rivers had this style of canoe, which was mainly utilised for small-scale fishing by traditional fishermen.

## Plank-built boat

Its common names include "Nao," "Lauka," and "Nauka." It is made of high-quality timber planks joined together with iron nails and has a spindle shape. To protect the bottom of the boat against borers and foulers, periodic coal and kerosene oil coatings are applied. The fishing boats range in size from 5-7 metres in length and 1-1.5 metres in width. While the larger fishing boats are utilised in rivers and reservoirs, the smaller boats employed in wetlands, lakes and tanks. This kind of boat is typically paddled manually by 2-4 fishermen.

**Fig. Dugout canoe**
*(https://g.co/kgs/9U4qUM)*

**Fig. Plank built boat**
*(https://mahasagarboats.com/stit)*

## Wooden framed iron/tin/aluminium sheet boat

Locally, wooden-framed ships with iron, tin, or aluminium sheets are called "Nauka" or "Dingi." Depending on their intended usage and the types of fishing to be done, these boats come in a variety of sizes and shapes. Only 1 or 2 fishers can be accommodated on typical boats between 3 and 5 metres long and 1.5 to 2 metres wide. These are mainly employed for low-volume fishing in shallow reservoirs.

## Large boat

To catch large-scale fish, large vessels are typically employed in group fishing. This type of boat requires more than two fishermen to operate and harvest the fish. The transportation of fish harvest also involves the employment of large vessels. It is primarily utilised in reservoirs and rivers.

**Fig. Large boat**

*(https://edition.cnn.com/travel/article/somnio-worlds-first-yacht-liner-unveiled/index.html)*

## Boat covered on the wooden frame and bamboo frame

Good-quality planks are used to construct some boats, which are then framed in wood or bamboo. Both the shape and size of these vessels vary. In rivers and reservoirs, they are mostly utilised for fishing. For running all kinds of equipment, this kind of ship is appropriate. Fishing can be done all year round on it, and because of its shades and ability to shelter users from the sun and rain, it also offers protection from the elements.

## Bamboo raft

This is the most basic kind of vessel used for transportation and fishing. The impoverished fishermen can readily make this with little expenditure because bamboo is widely accessible in this area. For this kind of craft, 5–6 bamboo should be enough. To build this raft, light bamboo is sliced horizontally into 15-20 identical pieces, each of the same size, and tied together with coir/jute rope (with all of the larger ends of the trunk facing the stern side). The lengths and dimensions of this kind of craft typically vary. The water flow and the amount of bamboo used mainly influence size. They are usually employed in slow-moving rivers and wetlands.

**Fig. Bamboo raft**

*(https://www.dreamstime.com/photos-images/bamboo-raft.html)*

**Fig. Wooden frames boat**

*(https://archive.shine.cn/feature/Reviving-ageold-bamboo-raft-tradition/shdaily.shtml)*

## Thermocol craft

It is a simple craft managed by a single person and primarily employed by poor fishermen. A larger thermocol box tied with rope is 0.7–0.8 metres long and 0.5 metres wide. In a different craft, bits of thermocouple are linked with a bamboo frame using string to create a bundle between 0.5 and 0.7 metres long and 0.4 and 0.5 meters wide. Fishermen used to sit atop these two bundles and travel to the water to fish.

## Rubber tube platform

It is the cheapest of all the crafts discovered so far; a single man handles it and is primarily used by impoverished fishermen. For this, old rubber tubes are utilized. Over the rubber tube is a platform made of wood, jute, or nylon fastened securely with rope. It is used to set and haul small mess gill nets in still and moving seas. Most often, it was found in rivers and reservoirs.

**Fig. Rubber tube platform**

*(http://ijrar.com/upload_issue/ijrar_issue_1397.pdf)*

## Fishing gears

The term "fishing gear" refers to all the tools and equipment used in the world's many fishing techniques, which use everything from bare hook and line to cutting-edge fish pumps. In general, there are two sorts of gears: active and passive. Thus, a vast range of fishing equipment is used, and these gears vary significantly in size, shape, fabrication material, operation mode, operating area, and type of fish to be caught. In Madhya Pradesh, fishing gear such as gill nets, cast nets, drag nets, scoop nets, handbag nets for jumping fish, hook and line and traps are used.

## Gill net

The gill nets are known locally as "Phansa Jal" or "Phanda Jal." Except for floating gill nets, it is a passive tool in which the mesh size changes depending on the body depth of the fish being pursued. The gill nets are kept vertically upright in the water column by flats and sinkers. The actual net meshes of this type of equipment have a considerable impact on catching certain fish species. This is one of the most popular types of gear being utilized in Indian open waters. It is employed in a wide range of depths, from shallow to deep. Gill nets with various mesh diameters were utilized throughout the state. The catch is either set using bamboo (for long-term operation) or released into the river by fishermen using boats to cross the river. The net is raised after a predetermined amount of time to retrieve the entangled fish. As floaters and sinkers, thermocol, light wooden pieces, plastic floats, and sinkers are widely used. The gill nets are normally 100-200 meters long and 2.5-10 metres wide. Monofilament gill nets were regularly employed instead of bulkier nylon nets.

**Fig. Gill net**

*(https://fishbio.com/gill-net-fishers-feel-the-pinch/)*

## Cast net

Cast net is commonly called "Phenka Jal" in the area. It has a conical shape with a thick rope tied to the tip, which, when extended out, forms a circle.

Several lead or iron weights are connected along the margin to sink it. Cast nets were frequently used in ponds, streams, rivers, reservoirs, and wetlands. In addition to using nets of varied mesh and pocket sizes, the length of the net is typically 2-2.5m, and the mesh size is 8-15mm to capture a certain species. Just one fisherman is running it. Cast nets are typically utilized year-round in quiet waterways. It can be used anywhere from the shoreline or with a craft in shallow water.

**Fig. Cast net**

*(https://en.wikipedia.org/wiki/Cast_net)*

## Drag net

Drag net is commonly called 'Maha Jal. This kind of net is towed through the water's surface in tiny or vast sheet of water. Its bulk, weight, and fine mesh (mesh size 20-50mm) necessitate the efforts of a sizable team of fishermen. The dragnet can be drawn manually or with a boat lowered into the sea. This net is used where the lake, wetland, or river bed is mainly quiet or stagnant. They operate in bays in reservoirs, especially in the summer when the water recedes, and fishermen avoid areas with submerged tree stumps.

**Fig. Drag net**

## Scoop net

Locally, the scoop net is referred to as "Penni." It consists of a bag net tied in the shape of a conical cone over a bamboo frame. One bamboo frame pole is stretched by roughly 1 m to run the net. The structure is submerged in water and then moved around the bottom for a distance before being raised. This equipment is often used in seas with a 0.5-2 metres depth. It is occasionally used year-round in various water bodies to catch small fish and prawns (Pravin *et al.*, 2011).

**Fig. Scoop net**

*(https://www.dreamstime.com/men-use-fish-net-to-scoop-water-image181918873)*

## Handbag net for jumping fish

"Jhalga" is used locally to describe this kind of bag net. It is a little handbag net with a square or rectangular shape that is fastened to a frame mode of cane, bamboo, or wood. It is mounted on a long handle used to operate the net. When fish jump out of the water, the square or rectangular frame covered in mosquito netting is utilized to catch them at a great distance. The handle's length varies depending on how far away the fishing site or area is, while the net's length ranges from 60 to 90 cm. It is typically operated alone in tumultuous water like waterfalls or other water bodies.

## Line and Hook

To catch different varieties of fish, different hooks and lines are employed. The lines can be secured from top to bottom, moored, drifted, or left in any position. Initially, a line was used to catch a single fish, but when it was made long, it was used to catch many fish at once. To ultimately hook the fish, real or fake bait is attached to the hook and used to entice the fish. Earthworms are typically used as live bait and snail flesh as dead bait in hook-and-line fishing. Hooks and bait are typically attached to the end of the fishing line to accomplish this. This gear is classified as 'Hand line' (with a smaller number of hooks) and 'Long line' (with a larger series of hooks).

## Traps

Traps are fishing equipment into which fish enter freely and become trapped. Trap fishing is a traditional passive fishing strategy. Its design, construction, and operation might vary widely depending on the target organism's location and behaviour. A trap may contain one or more compartments that close when fish enter them. Small traps, as opposed to large traps that remain open above the water's surface, are completely closed, like cages, except for the entrance. Bamboo splits are commonly used to build traps. In India, many types of fishing traps are employed (Manna *et al.*, 2011).

### V shaped trap

This kind of trap, known as a "Moude" locally, is used by one person to catch various fish. Winter to summer months saw a lot of "mode" trap fishing in shallow, still, and tranquil waters. This kind of trap frequently captures prawns and small fish.

### Long Trap

Locally, this kind of Trap is called a "Dhir." The Trap known as "Dhir" has a huge rectangular shape with several horizontal pockets. Using this Trap to catch fish in shallow, still, and tranquil waterways is typical from winter through summer. One or two fishermen used to run it. This Trap's long size makes it possible to harvest significant fish in a single operation. 'Dhir' can gather a variety of small fish species and prawns. In rivers, fish are caught with this kind of trap.

### Trap fitted with a net

'Chorya' is the local term for a Trap equipped with a net. This kind of Trap has a cylindrical shape with a nylon net that is between 5 and 10 metres wide and 10-15 metres long. The entire season can be employed with this trap. It is placed in the middle of the river where water is moving.

### Cylindrical trap

This kind of trap called locally known as a "Kumman," is used by one person to catch various fish. Winter to summer months saw a lot of "Kumman" trap fishing in shallow, still, and tranquil waters. This kind of trap can catch small fish and prawns.

## References

Baruah D., Dutta A. and Pravin P. (2013). Traditional fish trapping devices and methods in the Brahmaputra Valley of Assam. *Indian J. Tradit.* Know., 12(1), 123-129.

Bhattacharjya B.K., Manna R.K. and Choudhury M. (2005). Fishing crafts and gear of northeast India. Bulletin No. 142, CIFRI, Barrackpore, Kolkata, 67.

Bose A.K., Bose R. and Gupta S. (2017). Kunche Jal' Fishing: A low-cost indigenous fishery in Ganga River basin. *Int. J. Fish. Aquat. Stud.*, 5(4), 87-90.

Bose A.K., Bose R., Gupta S. and Das A.K. (2017). Dovali, the tribal fishing practice in river Narmada, MP, India. *Int. Res. J. Biol. Sci.*, 6(9), 26-30.

Bose A.K., Ridhi S.G., Suresh V.R. and Das A.K. (2017). Indigenous fishing (Char Kanta): A traditional fisheries in River Tawa, Madhya Pradesh, India. *International Journal of Fisheries and Aquatic Studies*, 5(3), 265-268.

Chourey P., Meena D., Varma A. and Saxena G. (2014). Study on fishing craft and gears of Bhopal district, Madhya Pradesh, India. *Int. J. Theor. Appl. Sci.*, 6(2), 65-67

Das R.K. and Barat S. (2014). Fishing gears operated in lentic and lotic water bodies of Cooch Behar district, West Bengal, India. Indian J. Tradit. Know., 13(3), 619-625.

Hameed M.S. and Boopendranath M.R. (2000). Modern fishing gear technology. Daya Publication House, New Delhi, 186. ISBN: 9788170352235

Kokate, A.A., Bhosale, B.P., Metar, S.Y., Chogale, N.D., Pawar, R.A. and Nirmale, V.H. (2016). Indigenous fishing crafts and gears of Krishna River with respect to Sangli district of Maharashtra, India. *Int. J. Fish. Aquat. Stud.*, 4(6), 434-438.

Kolekar V. (2009). An account of observations on fishing gear and crafts of Mandovi estuary of Goa. J. *Inland Fish. Soc. India*, 41(1), 41-46.

Manna R.K., Das A.K., Rao D.S.K., Karthikeyan M. and Singh D.N. (2011). Fishing crafts and gear in river Krishna. *Indian J. Tradit Know*, 10(3), 491-497.

Pravin P., Meenakumari B., Baiju M., Barman J., Baruah D. and Kakati B. (2011). Fish trapping devices and methods in Assam. *Indian J. Fish.*, 58(2), 127-135.

Singh S.N. (2008). River Narmada, its environment and fisheries. Bulletin No 157, CIFRI, Barrackpore, Kolkata, 64.

Sreekrishna Y. and Shenoy L. (2001). Fishing gear and craft technology. Directorate of Information and Publications of Agriculture, Indian Council of Agriculture Research, New Delhi, 342.

Welcomme R.L. (1983). River basins. FAO Fish. Technical Paper No. 202, 60. 6.
Gurumayum S.D. and Choudhury M. (2009). Fishing methods in the rivers of Northeast India. *Indian J. Tradit. Know.* 8(2), 237-241.

CHAPTER

# Impact of Environmental Factors on Inland Fisheries 13

## Introduction

Inland fisheries differ from marine fisheries because of their unique characteristics and the hydro-climatic factors that shape their specific and seasonal distribution. While some inland fisheries are highly commercialised, these fisheries are often typified by modest or domestic activity. As a result of this climatic fluctuation, also known as climate variability, many aquatic ecosystems experience significant seasonal, yearly, inter-annual, and decadal variations in the productivity and catchability of fish. The statistical distribution of weather patterns is significantly and persistently altered by environmental change across timescales ranging from decades to millions of years. The survival, growth, reproduction, and dispersion of individuals within a species are impacted by changes in average weather or the distribution of weather around those parameters. A population, a community, or an entire ecosystem may also experience effects (Brander, 2007).

The Earth's ecology has changed during the last few decades at both the global and local levels. An anonymous source summarised climate change in India in 2004. Among the changes that have an impact on inland fisheries are: (i) a national average increase in surface air temperatures of 0.4°C over the past century; (ii) An increasing pattern of monsoon season precipitation along the western coastline, in the northern region of Andhra Pradesh, and in northwestern India (with a range of +10% to +12% above the historical average over the past century), accompanied by a decrease in monsoon season rainfall; (iii) A heightened occurrence of severe storms overall, particularly along the coastal areas of Gujarat and West Bengal, at a rate of 0.011 incidents annually; (iv) A growing tendency toward heavy rainfall events; (v) A rise in sea levels ranging from 1.06 to 1.75 mm per year, consistent with the global estimate of 1-2 mm per year for sea level increase, as indicated (Unnikrishnan and Shankar 2007).

The aggregate catchment area of India's 12 main rivers is 252.8 X 106 hectares. 48 additional rivers are designated as medium rivers, with a combined catchment size of 24.9 X 106 hectares. Water discharged annually into the nation's rivers, which flow through various regions, totals 1869 km. Although some of these rivers are seasonal, most are open yearly. This is because a substantial portion of India experiences highly concentrated rainfall during two distinct periods: the summer monsoon season spanning from June to September/October and the tropical storm season extending from May to October, as noted by (Kale, 2003). The importance of water inputs from snow and glacier melt originating in high mountain regions is underscored by the fact that the Himalayan river systems in India, such as the Ganga and Brahmaputra, collectively produce approximately double the amount of water compared to the river systems in the southern peninsula, including the Godavari and Cauvery. The Indus River experiences an average rate of mountain glaciation of 3.4%, the Ganges River of 3.2%, and the entire Brahmaputra River of 1.3%. However, the most intense glacial activity is found in the tributaries of these river basins (Mall *et al.*, 2006).

The Gangetic Plains, encompassing the Main River and neighboring waterways, predominantly situated within the Indian states of Uttar Pradesh, Bihar, and West Bengal, form the basis for a diverse blend of artisanal, subsistence, traditional, and semi-intensive aquaculture practices that constitute the expansive Gangetic River system, the largest in India. The Gangetic River system has two hundred sixty-five freshwater fish species. Potential climate change scenarios suggest that the water challenges within the Ganga River basin could intensify, greatly impacting the ecosystem services and resources provided by inland water bodies, primarily fisheries. Striking a balance between the requirements of the aquatic ecosystem and other stakeholders in numerous marine environments across India could prove to be a complex task, given the rising population and associated demands, as highlighted by Das in 2009. Inland aquaculture and the production of capture fisheries are notably dependent on hydro-climatic factors. Fish are frequently ready to spawn when recent heavy rains and the water levels in interior reservoirs have started to rise. The management of sustainably managed inland fisheries is impacted by a significant fluctuation in the seasonality of hydro-climatic components in a given year.

## Causes of Environment Change

The elements that lead to changes in climatic conditions have been thoroughly examined by a number of writers, most notably Mackenzie and Mackenzie (1995), UNEP (2002), and IPCC (2007). They now know things like:

### Human Activities

On a worldwide scale, a prominent concern revolves around the release of greenhouse gases into the air as a result of human actions like the combustion of fossil fuels, gas flaring, coal utilization, and primarily deforestation as part of land use practices. The ongoing accumulation of these gases in the atmosphere has been a driving force behind the phenomenon of global warming. This term denotes the gradual rise in the Earth's overall temperature resulting from the emissions of greenhouse gases, and it is also linked to the immediate consequence of the depletion of the ozone layer, which arises from the Earth's surface absorbing ultraviolet (UV) light.

### Other Anthropogenic Activities

These include the development of nuclear and atomic weapons and other industrial pollutants humans brought on. These alterations could potentially amplify the quantity of solar radiation received by the Earth, thereby directly intensifying the impact of orbital parameters on the planet's radiation balance.

### Emission of Gases due to Volcanic Eruption

Due to ozone layer depletion and changes in the Earth's orbital characteristics, gases emitted during volcanic eruptions condense in the atmosphere, leading to global warming and a subsequent rise in the amount of solar radiation reaching the Earth.

## Impact of Environment change

### Rainfall & Monsoon

Precipitation, river inflow, snowmelt, reservoir levels, and other inland water bodies are some of the main hydro-climatic factors affecting inland waterways. In addition to numerous other hydrological occurrences, rainfall is the primary hydrological component that affects runoff, river input, and lake level (Chifamba, 2000). Numerous other elements, such as increases in atmospheric temperature (associated with global warming), cloud cover, evaporation, and more, influence these hydro-climatic factors both directly and indirectly. Specifically, human activities like irrigation, hydroelectric power generation, dam construction, and others have a significant impact on the hydrological characteristics of inland water bodies, including lakes and reservoirs. According to Moses *et al.* (2002), seasonal variation in hydro-climatic characteristics tends to impact the pool's output and catch.

Environmental change forecasts state that fish production will likely be considerably influenced, which may affect the economies of many developing nations worldwide (Allison *et al.*, 2009). Environmental change has an impact on economically essential fisheries both directly and indirectly. Direct consequences resulted from the fact that evaporation was higher and precipitation was lower. Indirect effects could result from using more water for irrigation to make up for lesser rainfall. Direct effects that affect physiology and behaviour change the fish community's growth, development, capacity for reproduction, mortality, and distribution. Fish are dependent on their habitats for protection and food, and indirect impacts change the productivity, structure, and composition of the ecosystems.

Variations in rainfall and water management pose additional hazards to inland fisheries. While some changes are predicted to improve fish production (Arctic Council, 2005), others have the reverse effect, diminishing reproductive capacity and rendering stocks more vulnerable to fishing pressure than was previously sustainable (Brander *et al.*, 2004). Particularly in freshwater and diadromous species like salmon (Friedland *et al.*, 2003) and sturgeon (Reynolds *et al.*, 2005), local extinctions are taking place near the margins of existing ranges. The Pacific oyster *Crassostrea gigas* is one competitive and damaging species that may expand to new places as a result of climate change (Diederich *et al.*, 2005). Environmental change has been connected to the catastrophic extinction of numerous aquatic species.

Rainfall raises the water levels of inland bodies, promoting fish spawning, reproduction, and natural fish recruitment. The quantities of fish caught show the non-random distribution and abundance of fish species throughout the year owing to seasonal variations. The catch distribution among places with varying amounts of rainfall was found to have a substantial impact on each species. In tropical areas, fingerling stockings and the accompanying suspension of fishing operations typically occur during the wettest months. The site with heavy rainfall can occasionally vary due to environmental changes and irregular rainfall patterns.

# Temperature

## Impact on the water body

Due to their preferred temperature range, the temperature considerably impacts the biological activity and growth of creatures that may survive in rivers and lakes. Higher temperatures also reduce the maximum capacity of water to dissolve oxygen, resulting in oxygen stress when absorbing significant quantities of organic substances. When temperatures deviate too much from the ideal range, either dropping too low or rising too high, the population of the species diminishes to the point of extinction. Due to its influence on water chemistry and chemical reactions, temperature is a critical factor. Increased temperatures facilitate the dissolution of more minerals from the rocks in which water is found, particularly groundwater, enhancing its electrical conductivity.

When a gas, like oxygen, dissolves in water, the situation is reversed. Since warm water has less dissolved oxygen than cool water, certain aquatic animals might not be able to survive there. Some substances are more dangerous to marine life at higher temperatures. The temperature of the water is a key determinant of its quality. After all, neither physical nor chemical characteristics are relevant to temperature. Because the animals and plants that dwell there have adapted to thrive in their habitat, temperature is not a fundamental problem for aquatic life in natural environments. When a body of water's temperature fluctuates due to a natural occurrence or human activity, the fish start to get nervous.

The warm water may startle the stream's aquatic organisms, which could reduce the stream's water quality. Particularly in the tropics, where there is no winter cooling and mixing like in temperate locations, these temperature changes assist lakes in maintaining a stable water column. Greater stability resulted in reduced water circulation, preventing nutrients from moving back from the deep water to the surface of the lake. These nutrients would have otherwise supported the growth of algae. The food chain's final link in the food chain is the fish that are essential for trade. Many lakes' water layers "turn" when the seasons change.

In the summer, the higher strata of the lake warm up more than the lower ones. Because warm water is less thick than cold water, it floats on the lake's surface. However, throughout the winter, some lake surfaces can get extremely chilly. The lake 'turns' when hotter water replaces the colder surface water as it lowers to the lake's bottom. The deeper water, which is now warmer than the surface and has a more stable temperature throughout the year, causes this since the surface water is getting denser than it formerly was. The layers don't mix, so they each grow distinct physical and chemical traits. The types of aquatic life, pH, dissolved oxygen levels, and nutrient concentrations in the upper and lower layers can vary greatly.

It nearly seems as though there are two separate lakes. Since the bottom layer no longer has access to the atmosphere, the lake's main supply of oxygen, the oxygen profile frequently exhibits the most noticeable alteration. When the top water cools down again in the autumn and reaches a temperature about equal to that of the lower water, stratification is no longer present. Because the two water masses have similar densities, the wind can stir and combine them. A similar process may take place when the lake mixes in the spring and the cooler top waters warm to the temperature of the bottom waters. Because the sun can heat a bigger volume of water in a small lake than in a deep lake, where light intensity exponentially diminishes with increasing water depth, a small lake may warm up more quickly and to a more significant temperature.

The lake's temperature is influenced by its size, temperature, and rate of water evaporation. It is well known that dissolved oxygen is less readily available and soluble at higher temperatures. High temperatures make it necessary for fish and other aquatic organisms to consume more oxygen; respiration approximately doubles for every 10°C change in temperature.

As a result, the demand for oxygen rises when the supply of oxygen decreases. According to research, many hazardous substances become more soluble and potent as the temperature increases. Because higher temperatures encourage the growth of sewage fungus and the putrification of sludge deposits already present in the aqua body system, they are harmful to the intended fish life. Each fish or other creature has a maximum temperature that it can endure, even in the presence of enough dissolved oxygen. Temperature increases result in mortality. An increase in atmospheric temperature affects water's physical and chemical characteristics and elevates the temperature of large bodies of water, including oceans, seas, lakes, and ponds. Various factors, including the fish species, their former environment, oxygen availability, and the combined effects of other contaminants, influence the highest temperatures adult fish can endure. As a result of global warming, lakes experience regression and an increase in water temperature. Global warming is well documented to have caused significant water loss in the Ganges River in India, the Aral Sea in Asia, the Yellow River in China, Owens Lake in Northern America, Chad Lake in Africa, and Platte Lake in Central Europe.

## Impact on Fish

Compared to trout and salmon, which are more susceptible to high temperatures, goldfish, bass, and carp can withstand them better. But for adult fish, these temperatures are necessary. For eggs to spawn and hatch, far lower temperatures are required. Many species have temperature-dependent reproductive limits. This is the main distinction between fish that live in cold and warm water. Cold-water species such as salmon, trout, and grayling have a minimum breeding temperature of just a few degrees above freezing. Many warm-water species, however, must reach far higher temperatures before they can start reproducing. As a result, they can only live and breed in environments where high temperatures are maintained for long enough for early development.

Trout eggs fail to hatch when exposed to temperatures exceeding 14.4°C. Chinook salmon eggs flourish at 16°C but cannot survive at 18°C. Rocky Mountain Whitefish eggs are negatively affected by temperatures between 20°C and 21°C. Consequently, many fish species cannot successfully complete their life cycles unless the temperature remains 10-15°C below their median tolerance limit throughout the spawning and hatching phases. As was already said, the appropriate temperature for fish depends on how quickly they develop and reproduce. It is conceivable that warm water fish will replace cold water fish in a section of the stream when the temperature has increased by 5 to 10° C. Without any evident direct mortality, the features of the fish and aquatic life they support will change.

Furthermore, temperature plays a crucial role in the successful hatching and propagation of eggs, and it is expected to undergo changes. Rapid temperature fluctuations are anticipated to have detrimental effects on fish life, with shifts of 5°C or more likely posing a significant threat. For instance, rainbow trout would struggle to withstand a temperature shock of 11°C above their acclimatization temperature, which is set at 12°C. However, they could tolerate a sudden increase of 8°C above their acclimation temperature of 10°C. Elevated temperatures may also stimulate processes that undermine the aesthetic quality of water resources, including the decomposition of sludge, the generation of sludge gas, the growth of saprophytic bacteria and fungi, and increased oxygen consumption due to decomposition processes. As the water rises, the body temperatures of cold-blooded fish rise, and they must sustain these temperatures by having faster metabolisms. In warm water, they need more food and oxygen to survive. Problems occur when fish cannot acquire as much oxygen due to rising temperatures. It is possible that fish do not receive the required amount of oxygen. As the seas warm, lethally low oxygen levels could cause many fish deaths. However, warm-water fish grow smaller bodies by 90% compared to their cold-water counterparts. Warm-water fish mature sexually more quickly than cold-water fish. The results are problems with reproduction and a decline in progeny.

Guppies produce fewer offspring in warm water, and grass carp ovulate less frequently. Salmon, catfish, and sturgeon need temperatures to drop below a certain point to spawn. As the water temperature rises, the diversity and amount of fish will decrease. When the water gets warmer, fish move to colder areas. If they go into deeper, farther-off seas, animals that depend on them for sustenance may starve. Due to rising temperatures in the Gulf of Alaska, fish have started moving to deeper waters, which led to the tragic loss of 120,000 seabirds in 1993, as they were unable to access their prey. Consequently, warm-water fish populations will gradually replace the cold-water species that previously inhabited these areas.

If lake temperatures increase by a mere 2°C, it is expected that warm-water species will outnumber cold-water species by a factor of 14. This influx of new species will engage in competition with native fish for resources and living spaces, potentially leading to reduced diversity within fish populations. These shifts in species composition, seasonality, productivity, and interactions among planktonic communities and their food chains are likely to impact water quality. Alterations in freshwater flow will also have repercussions on aquatic life, including changes in salinity, sediment input, and nutrient levels. Additionally, heightened thermal stratification can disrupt entire ecosystems by diminishing oxygen availability, damaging habitats, reducing biodiversity, and reshuffling the distribution of species.

The effects of global warming on aquatic ecosystems include higher water temperatures, lower lake levels, more species extinctions, and a growth in the habitats used by disease vectors. As the extinction of one species immediately affects other species, the disruption of the natural food chain can have serious consequences. Lake water temperature stands out among these variables as a crucial influence, affecting the needs of fish species for reproduction as well as the appropriateness of their habitats as a whole. Fish in the early phases of development, such as larvae and juveniles, are particularly susceptible because they are extremely sensitive to changes in water temperature. Although a fish population may be able to endure temperature changes in its particular environment for a short while, it nonetheless maintains a delicate ecological balance.

In cases of gradual and moderate environmental changes falling within a specific temperature range, fish typically undertake migration. Temperature plays a vital role in governing essential physiological processes such as feeding, respiration, osmoregulation, growth, and reproduction. However, if individuals within a population cannot acclimate to abrupt and extreme temperature fluctuations, they may undergo diminished metabolic activities and substantial mortality. The decrease of lake water's pH due to a rise in atmospheric $CO_2$ is another effect of climate change. In a research investigation contrasting the impacts of elevated and low pH levels on spawning and larval development, it was observed that 99% of eggs successfully hatched in high pH environments, in contrast to a mere 39.5% in low pH conditions. Moreover, 85% of spawns yielded healthy juveniles in high pH environments, while only 1.2% did so in low pH environments.

## Wind Pattern

### Underwater light Environment

When the wind traverses across the expanse of lakes, it generates surface waves. Despite the intricate physics governing wave dynamics, three fundamental factors—depth (d), fetch, and wind velocity—are pivotal in understanding wave behavior within shallow lakes. The fetch denotes the distance along the lake where the wind consistently blows. In simple terms, when the fetch is longer, for a given wind speed, the waves become taller. This phenomenon is readily observable on windy days at any lake. Typically, the water along the shoreline remains calm or only slightly disturbed due to the shielding effect from the wind. However, as the fetch extends farther away from the shore, waves grow in size and height.

The environmental consequences of wind-induced mixing and the cycles of resuspension and sedimentation are multifaceted. They have a notable impact on the underwater environment, particularly in terms of the quantity of suspended particles in the water column. Resuspension plays a significant role in altering the underwater lighting conditions, with the behavior of suspended particles in response to light determined by their composition, size, and shape (as noted by Davies-Colley & Smith in 2001). For example, clay particles tend to scatter light, while dissolved organic substances tend to absorb it. In cases where waters are clear, and there is minimal resuspension, light can penetrate shallow lake bottoms, fostering the growth of macrophytes and phytoplankton within the water column. Lower visual range and increased light limitation for organisms that require light for photosynthesis are the two main effects of light attenuation brought on by resuspension on the biota.

## Nutrient release from the sediments

One of the well-examined consequences of wind-driven resuspension is the release of phosphorus from sediment into the water. Shallow lakes and deep lakes exhibit significant differences in this regard. In deep lakes, nutrient-rich particulate matter settles near the lakebed and becomes mineralized as it descends through the water column. Nutrients are only incorporated into the epilimnion during seasonal turnovers, which occur when the entire lake mixes, typically in the autumn and spring. According to Guy *et al.* (1994), this theory suggests that stratified lakes could potentially lose up to 50% of their total phosphorus during the summer. The small depth of lakes ensures that nutrients and sediment are promptly reintroduced to the water column. Nutrient loss is unlikely to occur unless the lake is flushed, silt is dug up, or restoration methods like floating macrophytes beds pump the nutrients.

## Direct effects of mixing on phytoplankton: horizontal and vertical entrainment

Understanding the water motions by wind is necessary to comprehend the horizontal and vertical distribution of phytoplankton in the water column. According to Reynolds (2006), the phytoplankton distribution is far from homogeneous and can differ significantly along vertical and horizontal axes. However, vertical and horizontal dispersion does not co-occur or use the exact mechanisms.

In exposed lakes, the action of the wind generates internal currents responsible for the movement of phytoplankton cells. Verhagen (1994) employed a sophisticated hydrodynamic transport model for a moderately deep lake (with a 10 km fetch and a depth of 10 meters) and determined that the lateral dispersion of phytoplankton in response to variations in wind speed can span several weeks. However, the vertical response typically occurs within a 24-hour timeframe. In the case of the vast but shallow Lake Taihu in China, it was observed that the distribution of the buoyant cyanobacterium *Microcystis aeruginosa,* which forms blooms, is predominantly influenced by surface wind drift and, to a lesser extent, by internal currents (as reported by Wu *et al.* in 2010). Wind drift plays a vital role for buoyant phytoplankton species residing on the water's surface.

The primary factors influencing the vertical distribution of phytoplankton are their buoyancy and the level of mixing in the water. The effects of water mixing are more intricate in shallow lakes compared to deep lakes. In shallow lakes, mixing, whether initiated by artificial means or by the wind, often affects the entire water column. This mixing can have several consequences; it may resuspend nutrients and bring settled algae back into the euphotic zone. Conversely, it can increase water turbidity, leading to reduced productivity and algae growth. In the absence of wind, sinking algae tend to accumulate on top of the sediment. Additionally, many phytoplankton species have the ability to temporarily enter a resting state in response to darkness.

### Indirect effects of mixing on phytoplankton

Considering the information presented earlier, it becomes evident that wind-induced mixing influences various facets of the ecological dynamics in shallow lakes through multiple mechanisms. These include the resuspension of sediment, the release of nutrients, and a direct influence on the recruitment and competition among phytoplankton. Hellström (1991) hypothesised that sediment resuspension-induced more significant light limitation could hinder phytoplankton development and reduce productivity. The release of nutrients, especially phosphorus, can boost phytoplanktonic growth when there has previously been a nutrient constraint. Finally, excessive turbidity-induced phytoplankton loss can be compensated for or even reversed by benthic recruitment.

## References


Allison E. H., Perry A. L., Badjeck M. C., Adger W. N., Brown K., Conway D., Halls A. S., Pilling G. M., Reynolds J. D., Andrew N. L. and Dulvy N. K. (2009). Vulnerability of National Economies to the Impacts of Climate Change on Fisheries *Fish and Fisheries.* 10: 173–196.

Anonymous, (2004). India's initial national communication (NATCOM) to the United Nations Framework Convention on Climate Change. Government of India.

Arctic Council (2005). Arctic Climate Impact Assessment. Cambridge Univ Press, Cambridge, UK.

Brander, K. M. (2007). Global fish production and climate change. *Proceedings of the national academy of sciences,* 104(50): 709-714.

Brander, K. and Mohn, R. (2004). Effect of the North Atlantic Oscillation on recruitment of Atlantic cod (Gadus morhua). *Canadian Journal of Fisheries and Aquatic Sciences,* 61(9): 1558-1564.

Chifamba, P.C. (2000). The relationship of temperature and hydrological factors to catch per unit effort, condition and size of the freshwater sardine, *Limnothrissa miodon* (Boulenger), in Lake Kariba Fisheries Research. 45 (03): 271-281.

Das, M. K. (2009). Impact of recent changes in weather on Inland Fisheries in India. In P.K. Aggarwal (Ed.), Global climate change and Indian Agriculture, Indian Council of Agricultural Research. New Delhi, India, pp. 101–103

Davies-Colley, R. J., and D. G. Smith. (2001). Turbidity, suspended sediment, and water clarity: a review. *J. Am. water Resour. Assoc.* 37: 1085–1101.

Diederich, S., Nehls, G., Van Beusekom, J.E.E. and Reise, K. (2005). *Helgoland Mar Res.* 59: pp. 97–106.

Friedland, K. D., Reddin, D. G., McMenemy, J. R. and Drinkwater, K. F. (2003) *Can. J. Fish Aquat. Sci.* 2003; 60: pp. 563–583.

Guy, M., Taylor, W. D. and Carter, J. C. H. (1994). Decline in total phosphorus in the surface waters of lakes during summer stratification, and its relationship to size distribution of particles and sedimentation. *Can. J. Fish. Aquat. Sci.* 51: 1330-1337.

Hellström, T. (1991). The effect of resuspension on algal production in a shallow lake. *Hydrobiologia.* 213: 183–190.

Kale, V.S. (2003). Geomorphic effects of monsoon floods on Indian rivers. *Nat. Hazards* 28(I), 65–84.

Mall, R.K., Gupta, A., Singh, R., Singh, R.S. and Rathore, L.S. (2006). Water resources and climate change: An Indian perspective. Current Science 90(12), 1610–1626.

Moses, B.S., Udoidiong, O.M., Okon, A.O. (2006). A statistical survey of the artisanal fisheries of south–eastern Nigeria and the influence of hydroclimatic factors on catch and resource productivity. Elsevier Science Direct. 2002;57(3):267–278.

Reynolds, C. S. (2006). The Ecology of Phytoplankton, Press, Cambridge University press.

Reynolds, J. D., Webb, T. J. and Hawkins, L. A. (2005). Life history and ecological correlates of extinction risk in European freshwater fishes. Can *J Fish Aquat Sci.* 62(04): 854–862.

Jain, S. Sharma, G. and Mathur, Y.P. (2013). Effects of temperature variations on fish in lakes. *International Journal of Engineering Research & Technology.* 2, 10.

Unnikrishnan, A. S. and Shankar, D. (2007). Are sea-level-rise trends along the coasts of the north Indian Ocean consistent with global estimates, *Global Planet Change* 57(3–4): 301–307.

Vass, K.K., Das, M.K., Srivastava, P.K. and Dey, S. (2009). Assessing the Impact of climate change on inland fisheries in River Ganga and its plains in India. *Aquat. Ecosyst. Health Mgmt.* 12 (2): 138–151.

Verhagen, J. H. G. (1994). Modeling phytoplankton patchiness under the influence of wind-driven currents in lakes. *Limnol. Oceanogr.* 39: 1551–1565.

Wu, X., Kong, F. Chen, Y. Qian, X. Zhang, L. Yu, Y. Zhang, M. and Xing, P. (2010). Horizontal distribution and transport processes of bloom-forming Microcystis in a large shallow lake (Taihu, China). *Limnol. - Ecol. Manag. Inl. Waters* 40: 8–15.

CHAPTER 14

# Status of Ornamental Fisheries in India

## Introduction

Some fishes' varied forms and fascinating beauty have captivated people since ancient times, leading to their classification as ornamental fish. In nations such as China and Japan, goldfish and koi carp have enjoyed a rich history of being treasured as exquisite ornamental fish. The tradition of housing fish in aquariums traces its origins back to the year 1805. It was during this time that the first public aquarium came into existence at Regent's Park in England, marking a significant milestone in the world of aquatic display. Following this groundbreaking achievement, the practice of aquarium keeping surged in popularity, leading to the establishment of a total of 45 display aquaria by the year 1928. Presently, there are over 500 public aquaria worldwide. However, the global market for ornamental fish used in public aquaria accounts for less than 1%, while hobbyists primarily dominate the remaining 99%. What initially started as a hobby has become a thriving commercial aqua business, effectively becoming an industry. The entire ornamental fish industry is worth US $15,000 million. Considering the country's abundance of diverse indigenous fish resources, there is immense potential for becoming a significant player and strong competitor in the international ornamental fish trade (Nair,2004).

## Common Freshwater Ornamental Fish

In India, there is a rich diversity of exotic ornamental fish species, totaling 288 in number. Among these, 261 species are known to be egg-laying, while the remaining 27 are classified as live-bearing. Interestingly, a significant portion of typical food fishes in India also enjoys popularity as ornamental fish due to their aesthetic appeal. These particular fish species are collectively referred to as

"Indian Aquarium Fish" by both ornamental fish traders and hobbyists. In the state of West Bengal, the local fish fauna is highly diverse and includes a wide array of small fishes that may not find favor in traditional farming practices but possess substantial potential in the global fish market. An exploration of these fish species in their natural habitats has revealed a remarkably diverse group with considerable export value. In totality, there have been documented records of 176 indigenous ornamental fish species originating from various freshwater bodies across North and South Bengal.

These species are distributed across 98 different genera, falling under 41 distinct families and ten separate orders. Notably, the largest representation among these fish species comes from the order Cypriniformes, comprising 76 species, followed by Siluriformes with 51 species, Perciformes with 30 species, Clupeiformes with 5 species, Cyprinodontiformes with 4 species, Anguilliformes with 3 species, Osteoglossiformes with 2 species, and Synbrachiformes with 2 species. Additionally, Gasterosteiformes, Tetraodontiformes, and other orders each contribute 1 species to this rich diversity. Delving deeper into the family-level distribution, the family Cyprinidae stands out with the highest number of species, totaling 56. Following closely, Sisoridae is represented by 17 species, Balitoridae by 13 species, Bagridae by 12 species, Schibeilidae by 8 species, Channidae by 6 species, and Clupeidae, Belontidae, Mastacembelidae, and Siluridae each have 4 species. Meanwhile, Gobidae and Mugilidae contribute 3 species each, and Ambassidae, Aplocheidae, Ariidae, Nandidae, Notopteridae, Olyridae, Psilorhyndae, and Synbracidae are each represented by 2 species. Finally, Amblycipitidae, Anabantidae, Angilidae, Belonidae, Centrpomidae, Chacidae, Claridae, Eleotridae, Engraulidae, Heteropneustidae, Hemiramphidae, Lobotidae, Moringuidae, Ophichthidae, Pangasidae, Plotocidae, Scatophagidae, Sygnathidae, Tetraponidae, and Tetraodontidae each contribute 1 species to this diverse assembly of indigenous ornamental fish in Bengal (Mhapatra *et al.*, 2007).

## Diversity of Indigenous Ornamental Fish in India.

West Bengal has gained recognition for its extensive variety of decorative fish species within the Indian subcontinent. This area boasts a wide array of ornamental fish, encompassing loach, barbs, danio, snakeheads, trout, rasbora, gourami, catfish, clown fish, algae eater, and goby. These aquatic species encompass both formally categorized and uncategorized aquarium fish. The smaller ornamental fish encompass various species, including *Botia dario, Danio dangila, Puntius shalynius*, and *Schistura reticulofasciatus*. These fish are well-suited for life in aquariums and can thrive throughout their entire existence within this controlled environment. Conversely, among the larger species of food fish, such as *Neolissocheilus hexagonolepis, Labeo gonius, Channa marulius, Bagarius bagarius,* and *Rita rita,* there is a noteworthy transformation during their juvenile phase. These larger fish are now also recognized as ornamental fish when they are in their early stages of development, and they fall into the category of non-classified ornamental fish (Sinha and Jamal, 2015).

The various characteristics of native ornamental fish species allow for their classification. Some examples include fish with beautiful colours like *Pseudecheneis sulcatus* and *Tetradon cutcutia,* fish with stripes and banding patterns like *Botia rostrata* and *Brachydanio rerio,* fish with chameleon-like behaviour such as *Badis badis* and *Puntius shalynius,* fish with jumping behaviour like *Esomus danricus* and *Chela laubuca,* fish with charming predatory habits like *Channa orientalis* and *Glossogobius giuris,* fish with calm demeanour like *Ctenops nobilis* and *Nandus nandus,* fish with transparent bodies like *Chanda mama* and *Pseudambassis ranga,* tiny fish such as *Danio dangila* and *Brachydanio rerio,* robust fish like *Anabas testudineus,* and fish equipped with suction abilities, such as *Garra gotyla gotyla* and *Garra mcllendi,* are part of the group. Additionally, certain larger species of fish like *Neolissocheilus hexagonolepis, Labeo gonius, Channa maurulius,* and *Rita rita* are now recognized as ornamental fish when they are in their early stages of development, and they are categorized as unclassified ornamental fish (Sinha and Jamal , 2015).

## Breeding of Ornamental Fish

Various fish groups employ distinct reproductive strategies. Comprehending the breeding methods specific to each species proves crucial for successful breeding initiatives. Broadly speaking, fish can be classified into two main categories: Egg layers and Livebearers. Within these overarching categories, individual species have developed unique mechanisms to ensure the survival of at least a portion of their progeny.

### Livebearers

Livebearers refer to a group of fish species known for giving birth to live offspring instead of laying eggs. Within the category of livebearers, there exist two distinct types: ovoviviparous and viviparous. Ovoviviparous livebearers produce eggs that develop and hatch internally within the female's body before birth. On the other hand, viviparous livebearers do not form eggs; instead, the young ones receive nourishment through an umbilical-like cord or from secretions provided by the female (Biswas *et al.*, 2015). Livebearers are recognized for their high reproductive rates and ease of breeding, making them commonly found and widely bred species in the aquatic realm.

## Spawning tank

The live-bearing fishes, renowned for their unique reproductive strategy of birthing live offspring, demonstrate exceptional ease of breeding within an aquarium environment. Notably, the primary obstacle commonly encountered pertains to safeguarding the progeny against the cannibalistic tendencies displayed by their parental figures. To overcome this predicament, various innovative mechanisms have been developed to separate newborns from their mothers immediately after delivery promptly. Among the available options, the most productive arrangement involves the utilization of a screen crafted from mosquito netting, which is reinforced by a framework constructed from either stainless steel or wood. This screen can be securely installed across the tank, effectively confining the female to one section while permitting unrestricted passage for the young.

To promote the well-being and survival of young livebearers, it is essential to consider a holistic approach, taking advantage of natural resources while implementing specific measures. Despite the existence of separation devices designed for this purpose, the most effective and environmentally friendly method involves creating a thriving environment for aquatic plants, which can serve as a haven for the offspring. Simultaneously, removing the mother from the tank is crucial as soon as possible. Various aquatic plants are highly suitable for nurturing young livebearers, such as *Myriophyllum, Ambulia, Nitella, Utricularia,* and other similar species. By ensuring an abundance of these plants, the young will have ample shelter and protection, increasing their chances of survival. Additionally, promptly removing the mother prevents potential harm or predation towards the offspring, enabling their growth and development in a secure environment.

## Raising the Fry of Livebearers

Livebearer offspring, when they are still in their early stages of life, are initially quite sizable and capable of consuming either dried or prepared sustenance right away. Nevertheless, solely depending on pre-packaged food can result in subpar growth. Achieving desirable development requires a balanced diet that includes both live and dried sustenance. Especially during their initial phases, providing live food is of paramount importance to ensure their optimal growth and development. As they grow, the importance of live food diminishes, although a good proportion of it still enhances their well-being. Optimal choices for the first meals can encompass micro worms, recently emerged brine shrimp, minced earthworms, Daphnia, just-hatched mosquito larvae, or minced white worms. Appropriate selections in the realm of dehydrated sustenance encompass finely pulverized dried shrimp, processed cereals, and powdered liver or egg.

## Egg layers

The majority of species found in aquariums typically follow a reproductive process in which they deposit eggs and subsequently fertilize them externally. These fish can be categorized into five groups based on their egg-laying and egg-handling methods. These groups include egg-scatterers, egg-depositors, egg-buriers, mouth brooders, and nest-builders.

## Egg-scatterers

These species scatter eggs to substrate, plants, or surface. Egg scatterers reproduce by spawning in pairs or groups, typically without providing any form of parental care, and in some cases, they may even consume their own eggs. This reproductive strategy is characterized by the production of a significant number of eggs. Both Characins and Cyprinids are known to employ this method of reproduction.

## Spawning tank

To accommodate egg scatterers, special measures are needed in the spawning tank. For non-sticky egg-laying fish like Barbs and Danios, set up the tank with marbles or netting above the bottom to catch the eggs. After spawning, take out either the eggs or the parents from the tank.

For Tetras that are known for their adhesive egg scattering behavior, it is advisable to employ an appropriate foundation within the breeding tank and introduce delicate, thin-leaved aquatic flora. The eggs will adhere to the leaves naturally. After spawning, remove the parents from the tank to protect the eggs. These strategies address the specific needs of different egg scatterers, ensuring their eggs' successful development and hatching (Christian and Kelly,2020).

## Egg-depositors

In this scenario, eggs are placed in different locations. They can be laid on solid surfaces like stones or plant leaves or among delicate plants like Java moss. Egg deposition falls into two categories: those with parental care and those without. Some cichlids and certain catfish species provide care for their eggs. However, most egg-laying organisms, including cyprinids, various catfish, and Killifish, do not show parental care. They lay their eggs on a surface and then abandon them, usually without consuming their eggs.

## Spawning tank

Parents may stay in the tank after spawning when caring for egg-depositing fish with parental care. Substrate spawners rely on particular locations for breeding, such as aquariums with transparent walls, large leafy plants, or smooth, flat stones. In contrast, certain species, including Discus and Angelfish, have a preference for vertical surfaces. On the other hand, cavity spawners need specific spots like horizontally placed flowerpots, coconut shells, or rocky caverns for their breeding activities. To ensure security, the tank should have live or plastic plants.

For egg depositors that don't care for their young. Using a tank with delicate and broad-leafed plants, Java Moss, or artificial spawning mops is recommended. After spawning, the parents or the plants holding the eggs should be removed. If the plants containing eggs are taken out of the tank, it becomes essential to introduce fresh plants into the tank in order to enhance the prospects of future spawning.

## Egg-buriers

The members of this particular category of aquatic creatures inhabit arid environments with dwindling water sources. The predominant species within this group is the annual Killifish, known for its unique reproductive strategy of depositing eggs in mud. These Killifish exhibit rapid maturation and promptly produce offspring before succumbing to the diminishing water levels. The eggs enter a state of dormancy, patiently awaiting the arrival of rain to trigger their hatching process.

## Spawning tank

When dealing with egg-burying species, it is strongly advised to utilize a peat-moss substrate due to its advantageous qualities. Once the spawning procedure reaches its conclusion, the utilized peat moss can be easily extracted and prudently preserved within a plastic container. Under these conditions, it can sustain its viability for a duration spanning several weeks to months, contingent upon the particular species in question. When the moment arrives for another spawning cycle, a fresh substrate of peat moss can be introduced into the tank. To commence the hatching process, the conserved peat should be immersed in soft water, establishing the requisite conditions for the eggs to hatch with success.

## Mouth-brooders

Mouth-brooders refer to a category of fish species that engage in a unique reproductive behavior by sheltering and protecting either their eggs or newly hatched larvae within their oral cavity. There are two types of mouth-brooders: ovophiles and larvophiles. Ovophiles lay eggs in a pit or nest and then suck them into their mouth. The eggs are large but in small numbers. The eggs are incubated within the mother's oral cavity, where they eventually hatch, and the juvenile fish remain sheltered there for a period of time. Ovophile mouth-brooders, such as cichlids and certain labyrinth fish, exhibit this fascinating reproductive behavior.

Larvophiles lay eggs on rocks or plants, guarding them until they hatch. Once hatched, the female gathers the fry in her mouth, caring for them. She continues caring for the fry in her mouth until they can feed independently. At that point, the fry is released from the mother's mouth.

## Spawning tank

**Ovophile mouth-brooders:** Fish carrying eggs in their mouths can breed in the main aquarium for protection but separate them from other fish due to aggression. Provide appropriate tank setup.

Fish that do not engage in the larvophile mouth-brooding behavior, wherein they protect their eggs in their mouths, should be segregated into a distinct breeding tank. Eggs are vulnerable on surfaces, so a controlled environment ensures the survival of eggs and fry.

## Nest-builders

**Phage version:** "Numerous aquatic species construct various nests to protect their eggs. These domiciles vary from a simple depression dug into the gravel to elaborate bubble sanctuaries fashioned through the utilization of saliva-enveloped bubbles. Gouramis, Anabantids, and specific catfish breeds are notably abundant within the ranks of these fish renowned for constructing such abodes. These fish engage in brood care, demonstrating their commitment to nurturing their offspring.

## Spawning tank

Builders of nests ought to receive appropriate nesting materials. In the case of bubble-nest constructors, it is advisable to furnish them with delicate, floating vegetation while ensuring that the aquarium maintains a tranquil water environment devoid of disruptive currents to safeguard their nests. For those species that construct nests within the substrate, it is recommended to supply them with fine gravel or sand.

## Raising the fry of Egg layers

When the eggs hatch, the recently hatched larvae go through a remarkable metamorphosis, exhibiting few similarities to the adult fish from which they originated. Instead, they display a noticeable yolk sac of vibrant yellow colour and exhibit limited swimming abilities. Initially, the larvae depend on the nourishment provided by the yolk sac, gradually consuming it until all the yolk is absorbed—subsequently, the fry transition to consuming food from external sources.

For small fish fry, their initial sustenance can be obtained from Infusoria, commonly known as "green water" or egg yolk. As the fry grows, their diet can incorporate more substantial food items, including white worms, *Daphnia* (also called water fleas), *Artemia nauplii* (commonly called brine shrimp), and crushed or powdered fish flakes. These food options are beneficial as initial nourishment, especially for slightly larger fry belonging to the cichlid family (Biswas *et al.*, 2015).

As fish mature and increase in size, their dietary preferences and capabilities also expand. Initially, they consume smaller food items, but as they develop, they can accommodate larger meals, such as brine shrimp, larger Daphnia, fish flakes, insect larvae, and finely chopped Tubifex worms. Feed the young fish several times daily to help them grow and stay healthy. Sort them by size occasionally to keep bigger fish from eating smaller ones and prevent cannibalism.

## Breeding and Culture of Gold Fish (Carassius auratus)

The common goldfish, scientifically known as *Carassius auratus*, is a versatile fish with an omnivorous diet. This species falls under the Cypriniformes order and belongs to the Cyprinidae family. Goldfish have a diverse palate, consuming both live and synthetic food sources. They exhibit a wide spectrum of colors, including red, orange, silver, black, brown, white, and several others. This fish boasts an impressive array of over 30 distinctive varieties, some of which include Oranda, Lionhead, Fantail, Telescope Eye, Bubble Eye, Albino, and Pearl Scale.

To ensure an optimal aquatic habitat, it is essential to monitor and manage specific water conditions. These key water parameters should be taken into account:

- **pH Level:** pH range between 7.0 and 8.0.
- **Temperature:** Maintain a temperature range of 25-30 degrees Celsius.
- **Dissolved Oxygen Level:** Keep the dissolved oxygen levels within the range of 5-7 ppm (parts per million).
- Dissolved Free Carbon Dioxide: Ideally, maintain dissolved free

carbon dioxide levels between 0-4 ppm.

- Total Alkalinity: Aim for a total alkalinity range of 80-100 ppm.

For breeding purposes, you may also consider using an electric aerator (pump) to increase the dissolved oxygen levels to the optimal range of 6-7 ppm. Additionally, conducting partial water exchanges by replacing 25-30% of the water in the breeding tank can be a meaningful practice to help maintain these vital water parameters.

Goldfish breeding can be conducted in "*Gamlas,*" clay or cement containers with 40 to 60 litres capacity or rectangular glass aquariums with 50 litres. The goldfish generally produce 500 to 700 eggs, with the exact number depending on the size of the fish. To ensure a successful breeding process, a sex ratio of 2:1 (male to female) is maintained. The eggs are usually laid during the nighttime hours. Fertilized eggs exhibit a semi-transparent, grayish appearance, whereas unfertilized eggs are transparent and white in color. Due to their sticky nature, the eggs require a suitable substrate, such as soft weeds, tiles, or corals, to attach to. The fertilized eggs undergo hatching approximately between 4 and 7 days, with the specific duration contingent upon the prevailing water temperature. Goldfish do not display parental care; the parents may consume the newly hatched fry. Thus, it is necessary to remove the parents from the breeding tank once the breeding process is complete. (Pandey and Mandal, 2017).

**Mature fish**

**The base for adhesive eggs**

**Removal of eggs**

**Egg on mop**

|  |  |
|---|---|
| **Spawning tank and fry in the rearing tank** | **Marketable size** |
| *(Source:https://www.google.com/search)* | |

## Sex Determination

When male goldfish encounter specific conditions, they may exhibit the development of white protrusions or tubercles on their operculum and pectoral fin. Furthermore, in males, the central ray of the pectoral fin presents a pronounced thickness along its edge, whereas in females, this edge tends to be thinner. The fins of males become more pointed, whereas, in females, they appear rounded. The vent, or the posterior opening, takes on a concave shape with a small space in males, while in females, it becomes convex with a larger opening. Furthermore, the abdomen of males is observed to be smaller in size, whereas in females it is larger.

# Breeding of Tiger Barb (Barbus tetrazona)

The Tiger barb, scientifically identified as *Barbus tetrazona*, is one of over 30 freshwater ornamental fish species that hold commercial importance in Indian waters. This species distinguishes itself through its resilience, liveliness, and minimal upkeep needs. With their striking large scales, vibrant hues, inclination to form schools, and simplicity of reproduction and maintenance, tiger barbs have become highly sought after in the aquarium industry. It is essential to mention that, globally, there have been 1,078 barb species documented, yet only 70 of them have gained commercial significance owing to their attractive colour patterns.

## Maturity of barb

The tiger barb, known for its four unique stripes, usually attains sexual maturity at a total body length of 20-30 millimeters (equivalent to 2-3 centimeters) or when it's approximately 6-7 weeks old. Despite lacking sexual dimorphism, male tiger barbs exhibit a vibrant red hue on their fin rays and snout. At the same time, females tend to have a more rounded abdominal region and slightly subdued colours. Typically, in tiger barbs, it is common for the females to surpass the males in size, reaching a maximum length of 7 centimeters with a body depth of 2 centimeters. When it comes to their mating behavior, all tiger barbs participate in a one-on-one arrangement, where a lone male forms a pair with a lone female. The male displays aggressive conduct, while the female adopts a submissive role during this mating process.

## Broodstock conditioning

Separate tank conditioning is vital for tiger barb seed production. Begin by gathering 2-3 cm tiger barbs from ponds or water sources. Sort them by size, identifying females by their full round abdomens and males by their vibrant red fin rays. Place the chosen brooders in separate tanks, which can be circular, square, or rectangular, with rectangular tanks being easier for brooder management. Aim for a stocking density of one fish per four liters of water.

Tanks must be equipped with adequate aeration and maintained in an appropriate conditioning environment, ensuring a daily water exchange rate of 20%. The separated fish undergo conditioning by being fed frozen blood or tubi fex worms of Artemia. They are also provided with high-quality flakes or prepared food two to three times daily for two weeks.

During the conditioning process, it is essential to adopt preventive measures to safeguard the brood stock from potential infections, primarily from wild tubifex worms. This necessitates exercising the highest level of caution and implementing essential cleaning protocols. Furthermore, ensuring optimal water quality is crucial throughout the conditioning period, as the fish's diet can potentially lead to water contamination. Neglecting to condition the fish adequately may result in a substantial decrease in successful synchronized spawning occurrences (Mahapatra *et al.*,2005).

## Spawning of Tiger Barb

In the typical reproductive process, female tiger barbs often opt for submerged aquatic plants or roots as their chosen surface for depositing their eggs. When the actual spawning event unfolds, the male fish tightly embraces the female using its fins, during which both eggs and sperm are simultaneously released onto the selected substrate. This ritual may endure for several hours, or until the entirety of the eggs has been discharged. On average, each female tiger barb is capable of producing approximately 300 eggs in a single spawning cycle. However, it's important to note that tiger barbs display a rather voracious appetite for their own eggs post-spawning, and as such, it becomes imperative to promptly remove the parental fish from the breeding environment, as advised by Vogt and Wermuth in 1961. The eggs produced during this process exhibit adhesive properties and exhibit negative buoyancy in freshwater. They possess an average diameter of approximately 1.18±0.05 mm. Under ideal conditions, where the water temperature falls within the range of 25º to 27º Celsius, these eggs are expected to hatch within a span of three days.

Tigar Barb
*(source: https://www.google.com/search)*

## Breeding of Angel Fish (Petrophyllum Scalare)

The breeding of angel fish has undergone significant advancements, evolving into an art form through the development of various finnage types, such as veil finnages and superveil finnages, as well as the emergence of numerous colour variations. Surprisingly, a wide array of distinct variations emerged from the common silver angel fish in their natural habitat.

**Angel Fish**
*(source: https://www.google.com/search)*

When it comes to identifying the sex of angel fish, it can be challenging. However, during spawning, the reliable method for sex determination is observing the genital papillae. "These tiny protuberances bear a resemblance to miniature nipple-shaped protrusions and are commonly known as ovipositors. The female angel fish possesses a larger and more blunt ovipositor, while the male has a slimmer and more pointed one. These protrusions, which become visible at the vent area, serve the respective purposes of depositing eggs by the female and fertilizing them by the male (Mahapatra *et al.*, 2003)

## Spawning tank

A spawning tank with a generous capacity of 80 to 100 liters is the perfect choice. To create an ideal environment, ensure that the water within the spawning tank maintains a slightly acidic pH level, falling within the range of 6.6 to 6.8. Although it's possible for fish to spawn in conditions with a higher pH of 8, they exhibit a greater propensity to spawn when the pH levels are on the lower end of the spectrum, as mentioned earlier. If your intention is to keep the eggs alongside the parent fish, it becomes imperative to sustain these acidic water conditions. Maintaining a consistent pH range of 6.6 to 6.8 during the hatching process is essential, as it provides the optimal pH conditions necessary for the successful hatching of the eggs. Within the tank, it's beneficial to incorporate slates or slanted glass plates positioned at an angle. These structures serve to facilitate the spawning process by providing a suitable surface for the fish to lay their eggs. Additionally, placing an air stone in one of the corners of the tank will give mild aeration.

During spawning, the pair of fish meticulously selects a suitable location and diligently cleanses it approximately two to three days before the actual spawning event. The female fish, in particular, undergoes a series of trial runs. She situates her ventral fins or antennae in close proximity to the underside of her abdomen and arranges her anal fin in such a way that her lower body seems to maintain a relatively linear alignment. This particular positioning facilitates full engagement of her ovipositor with the selected surface, whether it's a slab of slate, a glass plate, or any other suitable substrate for the purpose of spawning. Following this, the male fish also performs a few practice runs. During the spawning process, as the moment approaches, the female gracefully glides across the chosen location and carefully deposits her eggs, which firmly adhere to the surface. Following this, the male approaches and swims alongside the freshly laid string of eggs, ensuring fertilization by releasing his reproductive fluids. To optimize the chances of successful fertilization, the male aligns his fins like the female's, allowing him to press against the eggs intimately. The male and female angelfish alternate their roles as they glide gracefully above the breeding area, engaging in this routine until they've deposited a substantial number of eggs, sometimes reaching several hundred or even more. The precise count of eggs is influenced by the female's size and overall health prior to the spawning event. Over the entire spawning duration, both parents remain dedicated, suspended in close proximity to the eggs, ceaselessly using their pectoral fins to create a delicate current that envelops and bathes the precious eggs. Within hours, some fertilized eggs may turn white, indicating that they are no longer viable and will subsequently be removed by the diligent parents (Mahapatra *et al.*, 2003).

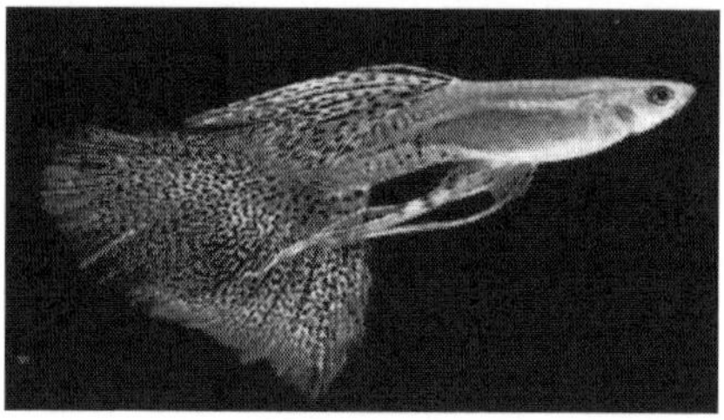

**Guppy**

**Molly**

**Southern platyfish**

**Green Sword tail**

**Kissing Gourami**

**Chocolate gourami**

*(Source: https://www.google.com/search)*

## Hatching Eggs

To ensure the successful hatching of the eggs, it is advisable to use soft water, with a preference for rainwater or distilled water. These sources naturally have a low pH, typically ranging from 6.2 to 6.5. Once the spawning process has been completed, transfer the glass plate containing the eggs from the spawning tank to a larger tank, preferably with a capacity of 30-50 liters, and make sure it is equipped with a sponge filter. To create a suitable environment for hatching, place a piece of slate against one of the tank's side walls. For proper oxygenation during the hatching process, it's essential to position an air stone in the tank in such a way that the air bubbles do not disturb the eggs excessively. To prevent fungal attacks on the eggs, add drops of Methylene blue, approximately 10% of the volume, to the tank. The hatching process typically spans a duration of approximately 36-48 hours, which can vary depending on the water temperature. After hatching, the young fry will tend to clump together before they start swimming freely. During this crucial period, it's essential to increase aeration in the tank to ensure that all the fry have access to an adequate supply of oxygen.

It is crucial to refrain from introducing any food into the tank until the fry have achieved the ability to swim freely. Typically, this milestone is reached within a period of approximately 3 to 5 days. Once the fry exhibit the capability to swim independently, you can begin introducing newly hatched brine shrimp as their primary source of sustenance. For rearing tanks housing baby angels that have surpassed the age of two weeks, it is acceptable to utilize standard dechlorinated tap water. It is advisable to perform daily water changes, replacing ten liters of water for every hundred liters present in the tank. These water changes should focus on removing water from the tank's bottom area where detritus tends to accumulate. It's worth noting that these rearing tanks do not necessitate any treatment to lower acidity levels (Mahapatra *et al.*, 2014).

## Feeding Schedule

In the initial one to three weeks of their development, juvenile Angel fish do not require feeding until they attain the stage of independent swimming. The duration of this phase may fluctuate, spanning anywhere between four to six days, contingent on the prevailing temperature conditions. Once the juvenile fry achieve free-swimming capability, it is advisable to provide them with freshly hatched brine shrimp (Artemia) or Moina as their primary source of nourishment. Initially, the brine shrimp should be directly provided to the young ones to ensure no excess floating in the tank for long periods. It is recommended to feed them three or four times a day. If any brine shrimp remains floating after 20 minutes, they have been overfed. Remember that providing small quantities helps prevent overfeeding and associated issues such as ammonia buildup and diseases (Biswas *et al.*, 2015).

### Three weeks to five weeks

Three weeks into their development, the juvenile fish reach a stage at which they are ready to consume finely crushed flake foods. These flake foods are offered sparingly as an additional dietary source. Following this initial three-week period, it becomes possible to introduce brine shrimp into their diet, with the fry readily consuming them within just 15 minutes of these tiny crustaceans being introduced into the aquarium.

### Five weeks to seven weeks

During the initial stage of their development, which occurs at the age of five weeks, the juvenile angel fish are gradually introduced to solid sustenance. To facilitate this transition, they are given a small amount of dry foods twice daily. This feeding routine continues until they reach the seventh week of their growth cycle. Throughout this particular period, the young angel fish consistently consume the dry flakes; however, they frequently expel them shortly after taking them into their mouths. While certain fish are willing to ingest the flakes, others lack interest. Nevertheless, as the seventh-week approaches, the angel fish gradually become more receptive to the dry flakes, and by this time, there should be minimal, if any, remnants of flakes left on the tank's bottom, unlike in the preceding weeks.

### Six weeks to adulthood

Once the young angel fish reach around six weeks old, they have developed to a point where they can begin consuming blended beef heart cubes. While baby brine shrimp can still be offered as food for the young angels for up to three months, providing them with beef liver and flakes alone is enough to facilitate their rapid growth.

## Breeding of Gouramies

While closely related to Bettas, gouramies distinguish themselves by lacking the aggressive tendencies found in their Betta counterparts. When placed in appropriate conditions, they exhibit a friendly disposition, making them suitable for community aquariums. One notable characteristic common to all gouramies is their pelvic fins, which take on the appearance of elongated, thread-like feelers, enabling versatile movement in all directions. Among the well-known gourami varieties favored by aquarium enthusiasts are the giant gourami (*Colisa fasciata*), dwarf gourami (*Colisa lalia*), pearl gourami (*Trichogaster leeri*), blue or three-spot gourami (*Treichogaster trichopterus*), moonlight gourami (*Trichogaster microcephalus*), snakehead gourami (*Trichogasrter pectolaris*), chocolate gourami

(*Sphaericthys osphronemoides*), and kissing gourami (*Helostoma temmincki*). A typical example of blue or three-spot gourami is presented above to describe the breeding of gouramies.

The three-spot gourami, a fish species, undergoes breeding activities from April to August. During this period, the mature male experiences a change in colouration, darkening in hue, while the female's abdomen becomes swollen. When selecting a breeding pair, choosing a mature female ready to spawn is crucial. This precaution is necessary due to the high aggression of male blue gouramis, which may harm or kill an unprepared female. Aquarium tanks with a capacity of 50-80 liters are suitable for breeding purposes. It is recommended to keep the water level in the tank below 25 cm to create the ideal conditions. To facilitate breeding, you can place one or two pieces of floating plants and beetle leaves on the water's surface. These will serve as platforms to support the bubble nest. There is no need for aeration in the tank.When pairing blue gouramis for breeding, it's important to maintain a balanced male-to-female ratio of 1:1. If the male is ready for breeding, it will typically construct a nest within one or two days. This nest takes the form of foam and floats just beneath the plant leaves. The male then entices the female under the leaves for courtship, during which the female releases a batch of approximately 20-25 eggs. The male diligently collects these eggs and attaches them to the floating nest (Das *et al.*, 2021).

## Culture of Ornamental Fish

Ornamental fish farming has become increasingly popular and profitable in India, offering an attractive alternative source of income. Its versatility sets this trade apart, as it can be pursued in rural and urban areas. The practice of cultivating ornamental fishes commonly involves the use of glass aquaria, concrete tanks, and net cages. In this industry, two primary approaches are commonly utilized for raising fish. The initial approach centers on the cultivation of juvenile fish from locally bred egg-laying species as well as live-bearing species. These species are nurtured by culturists interested in establishing their presence in the industry. Furthermore, indigenous farmers are actively involved in cultivating and nurturing imported juvenile specimens of exotic and valuable fish species, typically in return for monetary compensation. For individuals aspiring to venture into this sector, commencing with fish varieties like goldfish (*Carassius auratus*), platy (*Xiphophorus maculatus*), swordtail (*Xiphophorus helleri*), guppy (*Lebistes reticulatus*), and similar options presents a feasible opportunity. (Mohan, 2019).

## Feed Management

In ornamental fish farming, the types of feed play a crucial role and often incur high expenses. Selecting the appropriate feed in the right quantities is essential to ensure higher survival rates, optimal growth, and increased economic returns. Various species exhibit diverse dietary needs, underscoring the importance of acquiring in-depth understanding regarding the nutritional preferences and eating behaviors of the species that can be cultivated.

At present, hobby stores provide a range of feed options, but a considerable number of these choices come with high price tags. As a result, individuals interested in culturing can opt to create their own feed using readily accessible local resources, a solution that often proves to be more budget-friendly and cost-effective. Feeds typically classify into two primary categories: formulated dry feeds and live feeds.

## Live feed

In the aquarium industry, live feed plays a crucial role, as it significantly impacts the success of breeding and cultivating ornamental fish. The continuous availability of live feed is the primary determinant for the well-being of the larval stages of all aquarium fishes. Furthermore, even adult fish display a strong preference for live feed.

Live feed is a vital source of carotenoid pigments for enhancing ornamental fishes' vibrant colours. Consequently, the live feed is regarded as a living nutritious capsule encompassing all the necessary nutrients that contribute to improved breeding efficiency, exceptional growth, and enhanced fish colouration ( Mahapatra *et al.*,2005).

### Types of live feed

The various types of live feed utilized in the breeding and cultivation of ornamental fish are diverse. They encompass infusoria, Daphnia, tubificid worms, mosquito larvae, artemia, Moina, and more.

#### Infusoria

The infusorians, minuscule unicellular protozoans, can be found in virtually all aquatic environments. Following the absorption of the yolk sac, they serve as a superb initial nourishment source for recently hatched offspring. Among the numerous infusoria species, *Paramecium* and *Stylonychia* stand out as the frequently cultivated varieties for nourishing larvae. These organisms are predominantly observed in stagnant water sources containing decomposed kitchen refuse, emitting an unpleasant odor.

### Brachionus

Brachionus species serve as an ideal initial food source for larval fish due to their diminutive stature, leisurely swimming pace, tendency to remain suspended within the water column, and the feasibility of cultivating them in high-density environments.

### Tubificid worms

Tubifex worms are diminutive, crimson-colored creatures measuring up to 2 centimeters in length. They can be found in significant quantities within swiftly moving sewage drains. When these worms are disrupted or agitated, they swiftly retreat into the surrounding mud.

### Artemia

*Artemia,* a tiny aquatic organism found in saline environments, serves as a remarkable nutritional resource for both mature and young aquarium fish. The nauplii, or larvae of Artemia, are particularly beneficial. Artemia cysts, obtained from salt pans and commonly available in market tins, can be hatched in glass jars at 27-30 degrees Celsius. To prepare the hatching solution, 10-15 grams of salt is added to half a litre of boiled water, which is then allowed to cool. Aeration is provided to the water, and Artemia cysts are added at a rate of 0.5 to 1.0 grams per litre. The ideal pH range for hatching is 7.5-8.5. Hatching typically occurs within approximately 24 hours. Once hatched, the nauplii can be collected using their tendency to move towards the light (phototactic behaviour). A fine 100-mesh net is employed for the collection of nauplii, facilitated by the presence of 1000 lux lighting. Following the harvest, a meticulous washing process is carried out, and the nauplii are subsequently introduced into a receptacle filled with seawater featuring a salinity level ranging between 25 and 35 parts per thousand (ppt). These nauplii are subsequently fed to fish larvae based on their dietary requirements (Raja *et al.,* 2014).

## Formulated dry feeds

Additionally, it is essential to note that using probiotics in the feed can enhance ornamental fishes' growth and overall health. Probiotics such as lactobacilli and yeast can be included in the feed formulation. These probiotics aid in improving the digestion and absorption of nutrients and boosting the fish's immune system. To prepare the feed, a combination of the selected ingredients is used. The carbohydrates, proteins, and lipids essential for nutrition are readily supplied by vegetable sources such as groundnut oil cake, rice bran, tapioca flour, wheat bran, wheat flour, maize bran, and soybean meal. These ingredients are commonly available locally and are cost-effective.

Animal-derived sources like fish meal, silkworm pupae meal, prawn head meal, and earthworm meal are abundant in both proteins and vital amino acids. These ingredients contribute to the crude protein content required by the fish. Ensuring that these ingredients are of good quality and pathogens-free is crucial. The feed formulation for the ornamental fishes varies depending on their life stages. For young fishes, a feed with 40-50% protein, 4-6% lipid, and 40-50% carbohydrate is recommended. Adult fishes, on the other hand, require a feed with 30-35% protein, 6-8% lipid, and 40-50% carbohydrate. It is also recommended to incorporate 1% of both vitamins and minerals into the feed in order to fulfill the fish's nutritional needs.

Through the utilization of ingredients readily accessible within their local vicinity and adhering to the suggested feed formulations, individuals engaged in fish cultivation have the capability to craft high-quality feeds tailored to nourish their ornamental aquatic specimens. This approach reduces the dependence on expensive imported feeds and ensures that the fish receive the necessary nutrients for their growth and well-being (Mahapatra *et al.*, 2003).

## Health Care

Effective health management is crucial when it comes to ornamental fish farming. To ensure the well-being of the fish, certain measures are implemented. If tap water contains chlorine, it is left standing for one to two days to allow for dechlorination. In the case of pond water, purification is achieved by adding methylene blue at a rate of 3.5 mg per liter. To prevent the introduction of new pathogens, quarantine tanks are employed for newly acquired fish. In some cases, chemicals such as copper sulfate, potassium permanganate, malachite green, formaldehyde, and antibiotics like oxytetracycline or terramycin may be used as preventative measures against infections.

## References

Biswas, S. P., Singh, A. K., and Das, J. N. (2015). Conservation and management of ornamental fish resources of northeast India. *Journal of Aquaculture Research and Development*, 6(3).

Christian, R. L. and Kelly, M. (2020). Umatilla River Anadromous Fish Habitat Project.

Compendium, T. Freshwater *Fish Seed Production and Nursery Rearing in West Bengal, India.*

Das, A., Gujre, N., Devi, R. J. and Mitra, S. (2021). A review on traditional ecological knowledge and Its role in natural resources management: north east India, a cultural paradise. *Environmental management*, 1-22.

Mahapatra, B. K. and Lakra, W. S. (2014). Ornamental fishes of east Kolkata wetland, West Bengal, India. *Zoology*, 3(12), 406-408.

Mahapatra, B. K., Vinod, K. and Mandal, B.K. (2003). Studies on native ornamental fish of Meghalaya with a note on their cultural prospects. *Aquacult*4(2) : 173 – 182.

Mahapatra, B.K., Vinod, K. and Mandal, B.K. (2005). Export potentiality of native ornamental fish from North Eastern Hill States of India with a note for development of such fisheries. *Environ. Ecol.* 23(4): 780-786.

Mahapatra, B.K., Vinod, K. and Mandal, B.K. 2007. Native ornamental fisheries in Sikkim with its prospects and constraints.. *Environ. Ecol.* 25 (1): 125-128.

Mohan, S. K. (2019). Chemmattamvayal: The glorious wetland of Kasaragod district with a new record of Oriental Pratincole (*Glareola maldivarum*).

Nair K. S. (2004). Scope of expanding ornamental fish culture and trade and role of MPEDA. In: Ornamental fish culture and trade in Northeastern India (Eds. B. K. Bhattecharjya and M. Choudhury). Work Proc. CIFRI, Brackpore, Kolkata.

Pandey, P. K. and Mandal, S. C. (2017, May). Present status, challenges and scope of ornamental fish trade in India. In: Conference: Aqua Aquaria India, At Mangalore (pp. 1-10).

Raja, S., Babu, T. D., Nammalwar, P., Jacob, C. T. and Dinesh, K. P. B. (2014). Potential of ornamental fish culture and marketing strategies for future prospects in India. *International Journal of Biosciences and Nanosciences*, 1(5), 119-125.

Sinha, A., and Jamal, A. (2015). *Freshwater ornamental fish of Bihar. Central Inland Fisheries Research Institute.*

Vogt, D., and Wermuth, H. (1961). Knaurs Aquarien-und Terrarienbuch: das Haus-und Handbuch der Vivaristik. (*No Title*).

CHAPTER

# Wetlands & Their Importance

# 15

## Introduction

Wetlands are land areas where water covers the soil or remains near the surface for extended durations. They exist in diverse regions worldwide, including coastal areas, river floodplains, and even inland depressions. These unique ecosystems are characterized by the combination of water, hydrology, and vegetation that thrives in wet or waterlogged conditions. Wetlands can be classified into different types based on their specific characteristics, hydrology, vegetation, and geographical location. According to Ghermandi *et al.* (2008), wetlands are among the planet's most productive ecosystems. The human populations they serve also benefit greatly from their services (Ten Brink *et al.*, 2012). Despite this, they are adaptable and very sensitive to environmental changes (Turner *et al.*, 2000). These ecosystems vary greatly in terms of their origins, geographic locations, water regimes and chemistry, dominant species, and soil and sediment characteristics (Space Applications Centre, 2011). Wetland ecosystems are estimated to be valued over US$15 trillion yearly (MEA, 2005) and span an area of about 917 million hectares (m ha) to over 1275 m ha globally (Lehner and Döll, 2004). It is worth noting that wetland classification systems may differ depending on the source and purpose of classification.

## Hydrological Classification

- **Marine and Coastal Wetlands:** These wetlands occur along coastlines and are influenced by tidal movements and marine processes. They include estuaries, salt marshes, mangrove forests, and intertidal flats.
- **Riverine and Lacustrine Wetlands:** These wetlands are associated with rivers, streams, and lakes. They encompass floodplains, oxbow lakes, backwaters, and freshwater marshes.

## Vegetation-based Classification

- **Peatlands:** These wetlands are characterized by the accumulation of organic matter (peat) due to soggy conditions. Peatlands include bogs, fens, and mires.
- **Marshes:** Marshes are wetlands dominated by herbaceous vegetation such as grasses, sedges, and reeds. They can be freshwater or saltwater marshes.
- **Swamps:** Swamps are wetlands characterized by the prevalence of trees or shrubs. They can be freshwater swamps, saltwater swamps, or mangrove swamps.

## Geographical Classification

- **Coastal Wetlands:** Wetlands occurring along coastlines, including estuaries, mangroves, and salt marshes.
- **Inland Wetlands:** Wetlands are found away from the coast, encompassing floodplains, lakes, ponds, and peatlands.

## Ramsar Convention

The Ramsar Convention, also known as the Convention on Wetlands, is an international treaty established in 1971 in Ramsar, Iran. Its primary objective is to conserve and sustainably utilize wetlands designated as Ramsar sites, particularly those of international importance as waterfowl habitats. The convention classifies wetlands into nine main categories based on their ecological characteristics and functions: marine, coastal, riverine, lacustrine, palustrine, and human-made wetlands, with subtypes within each category (Ramsar Convention, 2016).

The core mission of the Ramsar Convention is to advocate for the conservation and wise use of wetlands worldwide. Floodplain wetlands, located in areas adjacent to rivers or other water bodies and experiencing periodic flooding as part of the natural hydrological cycle, hold significant importance. They contribute to regulating water flow, providing habitat for diverse species, and supporting the livelihoods of local communities. The Ramsar Convention plays a vital role in raising awareness about the ecological significance of these ecosystems and promoting their conservation and sustainable use globally. Through collaborative efforts, the convention contributes to protecting and responsibly managing floodplain wetlands, safeguarding their ecological functions, biodiversity, and benefits to human societies.

To ensure the protection of wetlands, the Ramsar Convention relies. The wetland conservation, as established in Article 2 and Article 3 of the convention. These articles provide the framework for the protection and sustainable management of wetlands.

## Article 2: Wise Use of Wetlands

Ramsar Convention, 1971, Article 2.1 specifies that member nations shall identify appropriate wetlands on their soil for inclusion in the List of Wetlands of International Importance (commonly known as Ramsar Sites) and guarantee their efficient management.

Article 2.2. Member nations must support the protection and judicious use of wetlands, taking into consideration ecological, economic, cultural, and scientific aspects.

## Article 3: List of Wetlands of International Importance

Article 3.1 establishes the List of Wetlands of International Importance and outlines the criteria for identifying and designating Ramsar Sites. Member countries are encouraged to select wetlands meeting the requirements (Ramsar Convention, 1971, Article 3.1).

Article 3.2 highlights the obligations of member countries to ensure the effective management and protection of Ramsar Sites, including monitoring, research, education, and information exchange (Ramsar Convention, 1971, Article 3.2).

## The function of the Ramsar convention

- **Recognition and Designation:** Floodplain wetlands are important ecologically, and the Ramsar Convention urges its signatory nations to locate and designate these wetlands as "Ramsar Sites." Ramsar Sites are areas that meet the convention's criteria for international importance, including ecological, botanical, zoological, hydrological, and cultural criteria.
- **Conservation and Wise Use:** Member countries must conserve and sustainably manage their Ramsar Sites, including floodplain wetlands. This involves taking measures to maintain the ecological character of the marsh, prevent degradation, and ensure the wise use of its resources. The wise use principle emphasizes the sustainable utilization of wetland resources while preserving ecological functions and biodiversity.

- **International Cooperation:** In order to conserve and manage floodplain wetlands, the Ramsar Convention encourages collaboration amongst its signatory nations.
- **Integrated Management:** The convention emphasizes adopting an integrated approach to managing floodplain wetlands. It involves considering wetland management's ecological, social, economic, and cultural aspects and involving various stakeholders, including local communities, indigenous peoples, and relevant government agencies.
- **Wetland Inventory and Monitoring:** Member countries are encouraged to develop inventories of their wetlands, including floodplain wetlands, and to monitor their ecological condition. Regular monitoring helps assess the effectiveness of conservation measures, detect changes in wetland health, and inform adaptive management strategies.
- **International Cooperation and Assistance:** The Ramsar Convention facilitates international cooperation and provides technical and financial assistance to member countries for wetland conservation and management. This support can help member countries address challenges related to floodplain wetlands, such as habitat degradation, water pollution, and conflicts between different land uses.

## Floodplain wetlands

India contains beels and other abandoned water bodies covering over 1.3 million hectares. In floodplains, "beels" refers to a variety of landforms with stagnant water, including oxbow lakes, sloughs, meander scroll depressions, residual channels, and back swamps. During floods or monsoon rains, beels shallow depressions linked to large rivers receive water from the rivers or catchment regions. In Assam, West Bengal, and Bihar, where many fishermen depend on them for a living, these water bodies are important fishing resources. A wide diversity of flora and fauna has developed as a result of the living communities' adaptation to the spatial and temporal variations inside the beels. Wetland ecosystems of various types may be found across India's numerous rural areas. Overview of the distribution of wetlands in India:

## 1. Coastal Wetlands

The Indian coastline has numerous coastal wetlands, including estuaries, lagoons, salt marshes, and mangrove forests. Notable coastal wetland areas include the Sundarbans in West Bengal, the Bhitarkanika Mangroves in Odisha, the Gulf of Mannar in Tamil Nadu, and the Rann of Kutch in Gujarat.

## 2. Riverine Wetlands:

India's extensive river network supports several riverine wetlands along their course, including floodplains and oxbow lakes. Examples of riverine wetlands include the Brahmaputra floodplains in Assam, the Gangetic floodplains in Uttar Pradesh and Bihar, and the Kolleru Lake in Andhra Pradesh.

## 3. Lakes and Reservoirs:

India has numerous lakes and reservoirs that serve as essential wetland ecosystems. Some prominent lakes and reservoirs include Dal Lake in J& K, Chilika Lake in Odisha, Vembanad Lake in Kerala, and Loktak Lake in Manipur.

## 4. High-altitude Wetlands

In the higher reaches of the Himalayas, wetlands are found in alpine and sub-alpine zones. Important high-altitude wetlands include Tsomoriri and Pangong Tso in Ladakh and Gurudongmar Lake in Sikkim.

## 5. Inland Wetlands

Inland wetlands, including marshes, swamps, and small ponds, are found across India. Inland wetlands include the Keoladeo National Park in Rajasthan, the Deepor Beel in Assam, and the Point Calimere Wildlife and Bird Sanctuary in Tamil Nadu.

# Oxbow lakes

Oxbow lakes are wetland formed when a meandering river or stream changes its course, leaving behind a curved or U-shaped body of water cut off from the main channel. Oxbow lakes are characterized by their distinctive shape and are important habitats for various aquatic species.

## Formation

- **Meandering River:** Oxbow lakes form due to a river's natural meandering (curving) over time.
- **Erosion and Deposition:** As a river meanders, erosion occurs on the outer banks (cut banks), and deposition occurs on the inner banks (point bars).
- **Cut-off Process:** Over time, the meander bend becomes more pronounced, and the river erodes through the narrow neck of the bend, creating a new, straighter channel.
- **Formation of Oxbow Lake:** The abandoned meander loop is filled with water, forming an oxbow lake separate from the main river channel.

## Characteristics

- **Shape:** Oxbow lakes have a curved or U-shaped appearance, resembling the bend of the original meandering river.
- **Water Source:** Oxbow lakes are typically fed by water from the surrounding area, such as runoff, groundwater, or occasional connections to the main river during flood events.
- **Hydrology:** They often have slower water flow than the main river, allowing sediments to settle and creating suitable conditions for wetland vegetation.
- **Biodiversity:** Oxbow lakes provide diverse habitats, supporting various aquatic and terrestrial plant and animal species. They can be particularly important for fish breeding and bird nesting.

## Ecological Importance

- **Habitat:** Oxbow lakes provide crucial habitats for many species, including fish, amphibians, reptiles, waterfowl, and various wetland plants.
- **Water Quality:** They serve as natural filters, helping to improve water quality by trapping sediments and nutrients and promoting nutrient cycling within the wetland ecosystem.
- **Flood Control:** Oxbow lakes can help mitigate flooding by absorbing excess water during high-flow events and providing additional storage capacity.
- **Biodiversity Conservation:** Due to their ecological significance, oxbow lakes are often protected as essential wetland areas and may be designated as Ramsar Sites or other conservation areas.

# Swamps

Swamps are wetland ecosystems characterized by the prevalence of trees or shrubs and soggy conditions. They are often found in low-lying areas with slow-moving or stagnant water. Here are some key features and characteristics of swamps:

## 1. Hydrology

Waterlogged Conditions: Swamps have permanently saturated or wet soils, resulting in poor drainage and water accumulation.

- **Water sources:** Swamps may receive water from precipitation, groundwater, nearby rivers or lakes, or a combination of these sources.

- **Slow Water Movement:** The water in swamps generally moves slowly, which allows for the accumulation of organic matter and the growth of wetland vegetation.

## 2. Vegetation

**Tree-Dominated:** Swamps are characterized by trees or tall shrubs that can withstand waterlogged conditions. Common tree species in swamps include cypress, willow, mangroves, and tupelo.

**Adaptations to Waterlogging:** The vegetation in swamps is adapted to the wet conditions, with some species having specialized root systems that can access oxygen in waterlogged soils.

**Canopy Structure:** The trees in swamps often form a dense canopy, providing shade and creating a unique habitat for various organisms.

## 3. Biodiversity

**Rich Habitat:** High levels of biodiversity are supported by swamps, which provide as homes for a variety of plant and animal species.

**Wildlife Diversity:** Swamps are home to diverse wildlife, including amphibians, reptiles, birds, mammals, and insects. They provide important breeding, nesting, and foraging grounds for many species.

## 4. Types of Swamps

- **Freshwater Swamps:** Found in freshwater environments, such as river floodplains and depressions. Examples include bottom land hardwood swamps and freshwater cypress swamps.
- **Saltwater Swamps:** Found in coastal areas where saltwater intrusion occurs. Mangrove swamps are a prominent example of saltwater swamps.

## 5. Ecological Functions

**Water Filtration:** Swamps act as natural filters, removing sediment and absorbing nutrients, improving water quality.

**Carbon Storage:** Swamps are significant carbon sinks, storing large amounts of organic matter as peat.

**Flood Control:** Swamps can help reduce the impact of flooding by absorbing excess water during periods of high precipitation or storm events.

Swamps play a vital role in the overall functioning of wetland ecosystems, providing habitat, supporting biodiversity, and delivering essential ecosystem services. They are valued for their ecological significance and are often protected and conserved as critical wetland areas.

## Marshes

Marshes are wetland ecosystems characterized by herbaceous vegetation, such as grasses, sedges, and reeds. They are transitional zones between land and water, with waterlogged or saturated soils. Some key features and characteristics of marshes:

### Hydrology

**Waterlogged or Saturated Soils:** Marshes have continuously saturated soils, often resulting in poor drainage.

**Shallow Water:** Marshes typically have shallow water depths, with water levels fluctuating seasonally or in response to rainfall and tidal influences.

**Nutrient-Rich Environment:** Marshes receive a regular supply of nutrients from the surrounding land and water, promoting the growth of wetland vegetation.

### Vegetation

- **Herbaceous Plants**: Marshes are dominated by herbaceous plants, including grasses, sedges, rushes, and reeds. These plants adapt to wet conditions and often have long, fibrous roots.
- **Emergent Vegetation**: The vegetation in marshes can grow above the water surface, forming a dense mat or cover that provides habitat and refuge for various organisms.
- **Zonation:** Marshes often exhibit zonation, with different plant species occurring in distinct zones according to their water depth and salinity tolerance.

### Biodiversity

**Habitat Diversity**: Marshes support a diverse array of plant and animal species due to availability of food, shelter, and breeding grounds.

**Avian Habitat**: Marshes are important for birds, providing nesting habitats, foraging areas, and stopover points during migration.

**Invertebrate Abundance**: Marshes are rich in invertebrate life, which serves as a food source for various species, including birds, amphibians, and fish.

## Types of Marshes

- **Freshwater Marshes**: Found in areas with freshwater sources such as rivers, lakes, or wetlands. Examples include prairie potholes, cattail marshes, and sawgrass marshes.
- **Saltwater Marshes**: Found in coastal regions influenced by tidal movements and saltwater intrusion. Salt marshes and mangrove swamps are common examples.

## Ecological Functions

- **Water Filtration**: Marshes act as natural filters, removing sediments and pollutants from water, thereby improving water quality.
- **Erosion Control**: The marshes' dense vegetation and root systems help stabilize shorelines and mitigate erosion by absorbing wave energy.
- **Carbon Sequestration**: Marshes effectively store carbon, with organic matter accumulating in the soil, contributing to climate change mitigation.

Marshes provide valuable ecosystem services, such as water purification, flood control, and habitat provision. They are ecologically significant habitats that support diverse plant and animal communities, contributing to the overall health and functioning of wetland ecosystems.

## Ramsar sites in India

**List of Ramsar sites in India (2023)**

*Source: https://www.pmfias.com/ramsar-sites-india*

| Sr. No | Ramsar Site | State | Designated Year | Area (km²) |
|---|---|---|---|---|
| 1 | Asan Barrage | Uttarakhand | 2020 | 4.44 km² |
| 2 | Ashtamudi Wetland | Kerala | 2002 | 614 km² |
| 3 | Bhakhira Wildlife Sanctuary | Uttar Pradesh | 2021 | 28.94 km² |
| 4 | Bhitarkanika Mangroves | Odisha | 2002 | 650 km² |
| 5 | Beas Conservation Reserve | Punjab | 2019 | 64 km² |
| 6 | Bhoj Wetland | Madhya Pradesh | 2002 | 32 km² |
| 7 | Chandra Taal | Himachal Pradesh | 2005 | 0.49 km² |
| 8 | Chilika Lake (Oldest Ramsar Site in India) | Odisha | 1981 | 1165 km² |
| 9 | Deepor Beel | Assam | 2002 | 40 km² |
| 10 | Gulf of Mannar Marine Biosphere Reserve | Tamil Nadu | 2022 | 526.72 km² |

| Sr. No | Ramsar Site | State | Designated Year | Area (km²) |
|---|---|---|---|---|
| 11 | Harike Wetland | Punjab | 1990 | 41 $km^2$ |
| 12 | Haiderpur Wetland | Uttar Pradesh | 2021 | 69 $km^2$ |
| 13 | Hokera Wetland | UT of Jammu and Kashmir | 2005 | 13.75 $km^2$ |
| 14 | Hygam Wetland Conservation Reserve | UT of Jammu and Kashmir | 2022 | 8.02 $km^2$ |
| 15 | Kanjirankulam Bird Sanctuary | Tamil Nadu | 2022 | 0.96 $km^2$ |
| 16 | Kanjli Wetland | Punjab | 2002 | 1.83 $km^2$ |
| 17 | Karikili Bird Sanctuary | Tamil Nadu | 2022 | 0.584 $km^2$ |
| 18 | Keoladeo National Park | Rajasthan | 1981 | 28.73 $km^2$ |
| 19 | Keshopur-Miani Community Reserve | Punjab | 2019 | 34 $km^2$ |
| 20 | Khijadia Wildlife Sanctuary | Gujarat | 2021 | 6 $km^2$ |
| 21 | Kolleru Lake | Andhra Pradesh | 2002 | 901 $km^2$ |
| 22 | Koonthankulam Bird Sanctuary | Tamil Nadu | 2021 | 0.72 $km^2$ |
| 23 | Loktak Lake | Manipur | 1990 | 266 $km^2$ |
| 24 | Lonar Lake (Impact Crater Lake) | Maharashtra | 2020 | 4.27 $km^2$ |
| 25 | Nalsarovar Bird Sanctuary | Gujarat | 2012 | 123 $km^2$ |
| 26 | Nanda Lake | Goa | 2022 | 0.42 $km^2$ |
| 27 | Nandur Madhameshwar | Maharashtra | 2019 | 14 $km^2$ |
| 28 | Nawabganj Bird Sanctuary | Uttar Pradesh | 2019 | 2 $km^2$ |
| 29 | Pala Wetland | Mizoram | 2021 | 18.5 $km^2$ |
| 30 | Pallikaranai Marsh Reserve Forest | Tamil Nadu | 2022 | 12.475 $km^2$ |
| 31 | Parvati Arga Bird Sanctuary | Uttar Pradesh | 2019 | 7 $km^2$ |
| 32 | Pichavaram Mangroves | Tamil Nadu | 2022 | 14.786 $km^2$ |
| 33 | Point Calimere Wildlife Sanctuary and Bird Sanctuary | Tamil Nadu | 2002 | 385 $km^2$ |
| 34 | Pong Dam Lake | Himachal Pradesh | 2002 | 156.62 $km^2$ |
| 35 | Ranganathittu Bird Sanctuary | Karnataka | 2022 | 5.18 $km^2$ |
| 36 | Renuka Lake | Himachal Pradesh | 2005 | 0.2 $km^2$ |
| 37 | Ropar Wetland | Punjab | 2002 | 13.65 $km^2$ |
| 38 | Rudrasagar Lake | Tripura | 2005 | 2.4 $km^2$ |

| Sr. No | Ramsar Site | State | Designated Year | Area (km²) |
|---|---|---|---|---|
| 39 | Sakhya Sagar | Madhya Pradesh | 2022 | 2.48 km² |
| 40 | Saman Bird Sanctuary | Uttar Pradesh | 2019 | 5 km² |
| 41 | Samaspur Bird Sanctuary | Uttar Pradesh | 2019 | 8 km² |
| 42 | Samaspur Bird Sanctuary | Uttar Pradesh | 2019 | 8 km² |
| 43 | Sandi Bird Sanctuary | Uttar Pradesh | 2019 | 3 km² |
| 44 | Satkosia Gorge | Odisha | 2021 | 981.97 km² |
| 45 | Shallbugh Wetland Conservation Reserve | UT of Jammu and Kashmir | 2022 | 16.75 km² |
| 46 | Suchindram Theroor Wetland Complex | Tamil Nadu | 2022 | 0.94 km² |
| 47 | Sultanpur National Park | Haryana | 2021 | 142.5 km² |
| 48 | Surinsar-Mansar Lakes | UT of Jammu and Kashmir | 2005 | 3.5 km² |
| 49 | Sur Sarovar (Keetham Lake) | Uttar Pradesh | 2020 | 4.31 km² |
| 50 | Sarsai Nawar Jheel | Uttar Pradesh | 2019 | 2 km² |
| 51 | Sundarban Wetland (Largest Ramsar Site in India) | West Bengal | 2019 | 4230 km² |
| 52 | Tampara Lake | Odisha | 2021 | 3 km² |
| 53 | Thane Creek | Maharashtra | 2022 | 65.21 km² |
| 54 | Thol Lake | Gujarat | 2021 | 6.99 km² |
| 55 | Tso Kar (High Altitude Ramsar Site) | UT of Ladakh | 2020 | 95.77 km² |
| 56 | Tsomoriri (High Altitude Ramsar Site) | UT of Ladakh | 2002 | 120 km² |
| 57 | Upper Ganga River (Brijghat to Narora) | Uttar Pradesh | 2005 | 265.9 km² |
| 58 | Vaduvur Bird Sanctuary | Tamil Nadu | 2022 | 1.12 km² |
| 59 | Vellode Bird Sanctuary | Tamil Nadu | 2022 | 0.77 km² |
| 60 | Vembannur Wetland Complex | Tamil Nadu | 2022 | 0.2 km² |
| 61 | Wadhvana Wetland | Gujarat | 2021 | 10.38 km² |
| 62 | Wular Lake | UT of Jammu and Kashmir | 1990 | 189 km² |
| 63 | Yashwant Sagar | Madhya Pradesh | 2022 | 8.22 km² |
| 64 | Asan Barrage | Uttarakhand | 2020 | 4.44 km² |
| 65 | Ashtamudi Wetland | Kerala | 2002 | 614 km² |
| 66 | Bhakhira Wildlife Sanctuary | Uttar Pradesh | 2021 | 28.94 km² |
| 67 | Bhitarkanika Mangroves | Odisha | 2002 | 650 km² |

| Sr. No | Ramsar Site | State | Designated Year | Area (km²) |
|---|---|---|---|---|
| 68 | Beas Conservation Reserve | Punjab | 2019 | 64 $km^2$ |
| 69 | Bhoj Wetland | Madhya Pradesh | 2002 | 32 $km^2$ |
| 70 | Chandra Taal | Himachal Pradesh | 2005 | 0.49 $km^2$ |
| 71 | Chilika Lake (Oldest Ramsar Site in India) | Odisha | 1981 | 1165 $km^2$ |
| 72 | Deepor Beel | Assam | 2002 | 40 $km^2$ |
| 73 | Gulf of Mannar Marine Biosphere Reserve | Tamil Nadu | 2022 | 526.72 $km^2$ |
| 74 | Harike Wetland | Punjab | 1990 | 41 $km^2$ |
| 75 | Haiderpur Wetland | Uttar Pradesh | 2021 | 69 $km^2$ |

## Wetland Services

The provided sentence describes the various ecosystem services provided by wetlands. These services can be categorized into four main types:

- **Provisioning services:** These include the immediate advantages of wetlands, such as the maintenance of fisheries, the production of peat for agriculture and energy, the harvesting of furbearers and other animals, the production of lumber, and the creation of natural foods.
- **Resgulating services:** Wetland ecosystems control and preserve environmental processes. In addition to improving water quality, they reduce river floods, safeguard coasts from cyclones, tsunamis, and other coastal storm surges, store carbon, and serve as habitat for rare and endangered animals.
- **Cultural services:** Wetlands have cultural significance and provide various cultural services. These include landscape aesthetics, serving as sites for human relaxation, facilitating ecology education, sustaining human cultures, supporting ecotourism, and providing opportunities for activities like bird-watching.
- **Supporting services:** These services are the underlying functions and processes that enable the other ecosystem services to occur. Wetlands serve as habitats for various species, including rare and endangered ones. They also serve as chemical sources, sinks and transformers, primary producers, water storage, and hydric soil growth.

Overall, wetlands are valuable ecosystems that provide a wide range of services, including direct resource provisioning, regulation of environmental processes, cultural benefits, and support for other ecological functions.

## Threats to Wetland Ecosystem

Freshwater wetland ecosystems, crucial for sustainability and well-being, have been extensively utilized, relied upon, and exploited (Molur *et al.*, 2011). In several areas, including North America, Europe, Australia, and New Zealand, over 50% of particular wetland types were transformed over the 20th century (MEA, 2005). Agricultural practises, dam construction, and other uses result in a yearly loss of wetland area of about 5000 km$^2$ in Asia (McAllister *et al.*, 2001). Worldwide, wetland ecosystems are under tremendous strain due to this strong reliance on water and resources, which has a negative impact on species diversity and populations (Molur *et al.*, 2011). As a result, many wetlands-dependent species, such as 21% of bird species, 37% of mammal species, and 20% of freshwater fish species, have either gone extinct or are now threatened globally (MEA, 2005).

Reduced wetland acreage has been shown to have detrimental effects on the fundamental functions that wetlands provide, often known as ecosystem goods and services (Zedler and Kercher, 2005). Urbanisation, changes in land use, drainage issues in agriculture, infrastructural growth, industrial and agricultural pollutants, and the effects of climate change and variability are the primary causes of wetland loss worldwide. The following sections will explore several factors responsible for notable transformations in India's wetland ecosystems.

## Pollution from Agriculture, Cities, and Industries

The introduction of pesticides and fertilisers from agricultural runoff and industrial and urban wastewater discharge has significantly worsened the water quality in many rivers, lakes, streams, and wetlands across Asia. Eutrophication has expanded widely as a result of these processes (Liu and Diamond 2005; Prasad *et al.* 2002).

### Other threats

According to the Millennium Ecosystem Assessment (MEA, 2005 and UNESCO (2007), the loss and alteration of wetland ecosystems are anticipated to be influenced significantly by global climate change. These findings are especially pertinent to the Indian subcontinent, where both the frequency and average temperature of the atmosphere have increased. Due to the increased atmospheric concentration of greenhouse gases such $CO_2$, $CH_4$, and $N_2O$, there have been fewer rainy days and less precipitation overall throughout time (Bates *et al.*, 2008).

## References

An Introduction to the Ramsar Convention on Wetlands, 7th ed. (previously The Ramsar Convention Manual). Ramsar Convention Secretariat, Gland, Switzerland.

Assessment, M. E. (2005). *Ecosystems and human well-being: wetlands and water*. World resources institute.

Bassi, N., Kumar, M. D., Sharma, A., & Pardha-Saradhi, P. (2014). Status of wetlands in India: A review of extent, ecosystem benefits, threats and management strategies. *Journal of Hydrology: Regional Studies*, 2, 1-19.

Bates, B.C., Z.W. Kundzewicz, S. Wu and J.P. Palutikof, Eds., 2008: Climate Change and Water. Technical Paper of the Intergovernmental Panel on Climate Change, IPCC Secretariat, Geneva, 210 pp.

CEBPOL, N. (2018). Mainstreaming Biodiversity: Inland Fisheries and Aquaculture-A key for food and nutritional Security, Published by the Centre for Biodiversity Policy and Law, National Biodiversity Authority, 42p.

Finlayson, C.M., Spiers, A.G. (Eds.), 1999. Global review of wetland resources and priorities for wetland inventory. Supervising Scientist Report 144/ *Wetlands International Publication* 53, Supervising Scientist Canberra.

Ghermandi, A., Van Den Bergh, J. C., Brander, L. M., De Groot, H. L., & Nunes, P. A. (2010). Values of natural and human-made wetlands: A meta-analysis. *Water Resources Research*, 46(12).

Koester, V. (2014). The Ramsar Convention: Which parties are effectively parties to what?(And other legal issues). *Envtl. Pol'y & L.*, 44, 100.

Lehner, B., & Döll, P. (2004). Development and validation of a global database of lakes, reservoirs and wetlands. *Journal of hydrology*, 296(1-4), 1-22.

Liu, J., & Diamond, J. (2005). China's environment in a globalizing world. *Nature*, 435(7046), 1179-1186.

McAllister, D. E., Craig, J. F., Davidson, N., Delany, S., & Seddon, M. (2001). Biodiversity impacts of large dams. *Background paper*, 1.

Meskell, L. (2014). States of conservation: protection, politics, and pacting within UNESCO's World Heritage Committee. *Anthropological Quarterly*, 217-243.

Molur, S., Smith, K. G., Daniel, B. A., & Darwall, W. R. T. (2011). The status and distribution of freshwater biodiversity in the Western Ghats, India. *Cambridge, UK and Gland, Switzerland: IUCN, and Coimbatore, India: Zoo Outreach Organisation.* ISBN: 978-2-8317-1381-6

Prasad, S. N., Ramachandra, T. V., Ahalya, N., Sengupta, T., Kumar, A., Tiwari, A. K., ... & Vijayan, L. (2002). Conservation of wetlands of India-a review. *Tropical Ecology*, 43(1), 173-186.

Ramsar Convention (1971). The Ramsar Convention on Wetlands: Text of the Convention. Retrieved from: https://www.ramsar.org/sites/default/files/documents/library/text-eng.pdf

Ten Brink, P., Badura, T., Farmer, A., & Russi, D. (2012). The economics of ecosystem and biodiversity for water and wetlands: a briefing note. *Institute for European Environmental Policy, London*, 9.

Turner, R. K., Van Den Bergh, J. C., Söderqvist, T., Barendregt, A., Van Der Straaten, J., Maltby, E., & Van Ierland, E. C. (2000). Ecological-economic analysis of wetlands: scientific integration for management and policy. *Ecological economics*, 35(1), 7-23.

Wetlands International. (2010). Oxbow lakes. Retrieved from: https://www.wetlands.org/our-work/what-we-do/understand/oxbow-lakes/

Zedler, J. B., & Kercher, S. (2005). Wetland resources: status, trends, ecosystem services, and restorability. *Annu. Rev. Environ. Resour.*, 30, 39-74.

CHAPTER

# Fisheries Management & Conservation Laws

16

## Introduction

The fisheries sector is uniquely positioned in India's national economy, socio-cultural context, rural employment, and food and nutritional security. Due to its extensive scope, favourable ecological conditions, and significant potential for growth, the sector continues to offer both direct and indirect livelihood opportunities to a substantial number of people. The sector's development objective is to attain self-sufficiency in fish production, improve the socio-economic conditions of fishing communities, ensure high-quality nutrition, generate foreign currency earnings, and alleviate poverty through the conservation and sustainable utilisation of fisheries resources. In addition to abundant natural perennial and seasonal water bodies, as well as artificial inland water bodies, the country possesses an 8129 km coastline, a 2.02 $km^2$ exclusive economic zone (EEZ), and vast areas of estuarine and brackish water that serve as breeding grounds for commercially valuable penaeid prawns and various species of finfish. Fishing has been a traditional occupation for inland and coastal fishing communities, whose livelihoods rely entirely on fishing activities.

A dominant approach to wildlife conservation has been the creation of protected areas where many human activities, especially hunting and fishing, are banned, restricted, or significantly regulated by law as precautions against their potentially negative impacts on wildlife (Bruner *et al.* 2001; Levering-ton *et al.* 2008; Kaur *et al.* 2012; Mitra and Bhaumik 2013; Velho *et al.* 2012; Ghosh-Harihar *et al.* 2019). Fisheries management and conservation are essential to ensure fishery resources' sustainable use and preservation. They involve various strategies and measures to regulate fishing activities, protect fish populations, and maintain aquatic ecosystems' overall health and productivity.

Diverse environments in India provide a home to a diverse range of aquatic life. Finfishes, shellfish, molluscs, crustaceans, and echinoderms are considered fish in this context, which refers to all aquatic species. Indian waters are home to around 9.2% of the 32,042 finfish species known globally. Along with finfish, India has a rich aquatic genetic resource base, including 765 species of echinoderms, roughly 2934 species of crustaceans, and 5070 species of molluscs. Due to the low level of domestication in the fishing industry, natural aquatic genetic resources are significant because most genetic resources for food still come from the wild.

- **Stock Assessment:** Fish stocks are regularly monitored and assessed to estimate their abundance, distribution, and biological characteristics. This information helps determine sustainable catch limits and identify any signs of overfishing or population decline.
- **Fishing Regulations:** Fisheries management involves implementing and enforcing fishing regulations to control fishing efforts and prevent overexploitation. These regulations may include fishing seasons, gear restrictions, size limits, and catch quotas. By ensuring that fishing practices are sustainable and responsible, these measures help maintain healthy fish populations.
- **Marine Protected Areas (MPAs):** Establishing MPAs is an effective conservation tool. These designated areas restrict or prohibit fishing activities to protect essential habitats, breeding grounds, and vulnerable species. MPAs contribute to biodiversity conservation, stock recovery, and the overall resilience of marine ecosystems.
- **Gear and Bycatch Reduction:** Selective fishing gear and techniques help reduce unwanted bycatch (non-target species) and minimise the impact on the marine environment. By utilising gear modifications, such as escape panels or mesh sizes, fishermen can improve selectivity and reduce unintentional catches of non-target species.
- **Collaborative Management:** Engaging stakeholders, including fishermen, local communities, scientists, and government agencies, is crucial for effective fisheries management. Collaborative approaches foster dialogue, knowledge sharing, and the incorporation of local ecological knowledge, leading to more informed and inclusive decision-making processes.

- **Ecosystem-Based Management:** Fisheries management is shifting towards an ecosystem-based approach, recognising the interdependencies and interactions within marine ecosystems. This approach considers not only the target species but also their habitats, predators, and prey, as well as broader ecological and environmental factors.
- **International Cooperation:** Many fish species migrate across national boundaries, making international cooperation essential for effective fisheries management and conservation. International agreements, such as regional fisheries management organisations (RFMOs), facilitate collaboration on shared fish stocks and ensure coordinated conservation efforts.
- **Research and Data Collection:** Continuous research and data collection are vital for understanding fish population dynamics, ecosystem health, and the impacts of fishing practices. This knowledge forms the basis for evidence-based decision-making and adaptive management strategies.

## International Union for Conservation of Nature (IUCN)

The International Union for Conservation of Nature (IUCN) maintains the IUCN Red List of Threatened Species, the world's most comprehensive inventory of the conservation status of species. The Red List categorises species into various threat categories based on their population size, trends, distribution, and threats they face. Here are the main categories used in the IUCN Red List:

- **Extinct** (EX): No longer found alive species.
- **Extinct in the Wild** (EW): Species are known only to survive in captivity or cultivation.
- **Critically Endangered** (CR): Species facing an extremely high risk of extinction in the wild.
- **Endangered** (EN): Species facing a very high risk of extinction in the wild.
- **Vulnerable** (VU): Species facing a high risk of extinction in the wild.
- **Near Threatened** (NT): Species close to qualifying for a threatened category but do not currently meet the criteria.
- **Least Concern** (LC): Species that are widespread and abundant and are not currently considered to be at risk of extinction.

- **Data Deficient** (DD): Species for which there is inadequate information to assess their conservation status.
- **Not Evaluated** (NE): Species that have not yet been assessed on the Red List.

## Conservation of Genetic Diversity

Conservation of genetic diversity is a critical aspect of overall biodiversity conservation. Genetic diversity refers to the variety of genetic information within a species or population, including genetic traits, variations, and adaptations. Conserving genetic diversity is essential for several reasons:

## Adaptability and Resilience

Genetic diversity allows populations to adapt and respond to changing environmental conditions, such as climate change or disease outbreaks. Greater genetic diversity provides a broader pool of genetic traits that can enhance a population's ability to survive and thrive in diverse environments.

## Species Survival and Evolution

Genetic diversity is essential for the long-term survival of a species. It helps maintain healthy populations, reduces the risk of inbreeding and associated genetic disorders, and facilitates evolutionary processes by providing the raw material for natural selection.

## Ecosystem Functioning

Genetic diversity plays a role in maintaining ecosystems' overall health and functioning. It supports ecological interactions, such as pollination, seed dispersal, and predator-prey relationships, contributing to ecosystem stability and resilience.

**For the conservation of genetic diversity, several approaches and strategies can be employed**

- **Protected Areas and Habitat Conservation:** Establishing protected areas and conserving natural habitats is crucial for preserving genetic diversity. By protecting intact ecosystems and providing sufficient habitat for species, we can ensure the maintenance of diverse gene pools.
- **Captive Breeding and *Ex Situ* Conservation:** In cases where species are critically endangered or face immediate threats, *ex-situ* conservation measures, such as captive breeding programs, can help preserve genetic diversity. These programs aim to maintain genetically

diverse populations in controlled environments and facilitate the reintroduction of individuals into the wild when suitable habitats are available.

- **Habitat Restoration and Connectivity:** Enhancing and restoring degraded habitats and establishing ecological corridors can promote gene flow between fragmented populations. This allows for the exchange of genetic material and helps maintain genetic diversity within and among people.
- **Sustainable Use and Harvesting Practices:** Implementing sustainable management practices for commercially exploited species helps prevent over harvesting and maintains genetic diversity. This may include setting appropriate harvest quotas, regulating fishing or hunting methods, and promoting sustainable agricultural practices.
- **Seed Banks and Genetic Resource Centres:** Seed banks and genetic resource centres collect and preserve genetic material from diverse plant species. These repositories serve as valuable genetic resources for future restoration efforts and research.

## Monitoring and Research

Regular monitoring of populations and genetic studies is essential to assess the status of genetic diversity and identify any threats or changes over time. This information guides conservation actions and helps prioritise species and populations needing protection.

Conservation of genetic diversity requires a combination of efforts at the species, population, and ecosystem levels, as well as collaboration among scientists, policymakers, local communities, and conservation organisations. By conserving genetic diversity, we can help safeguard species' and ecosystems' long-term viability and evolutionary potential.

### Conservation of Ecological Diversity

Conservation of ecological diversity, also known as ecosystem diversity, focuses on preserving the variety of ecosystems and their ecological processes on Earth. Ecological diversity encompasses the range of habitats, communities, and environmental interactions that exist within and between ecosystems. It is crucial for maintaining ecosystems' health, stability, and functionality.

- **Habitat Conservation:** Protecting and conserving a wide range of habitats is fundamental to maintaining ecological diversity. This includes terrestrial ecosystems like forests, grasslands, wetlands, and deserts and marine and freshwater habitats such as coral reefs, mangroves, rivers, and lakes.

- **Protected Areas**: Establishing and effectively managing protected areas, such as national parks, nature reserves, and wildlife sanctuaries, is vital in conserving ecological diversity. These areas serve as refuges for various species and ecosystems and provide essential habitats for migratory species.
- **Ecological Restoration:** Restoring degraded or destroyed ecosystems is essential for conserving ecological diversity. Efforts may involve reforestation, wetland restoration, habitat reconstruction, and reintroduction of native species. Restoration enhances ecosystem functionality, biodiversity, and the overall resilience of ecosystems.
- **Sustainable Land and Water Management:** Adopting sustainable land and water management practices is crucial for conserving ecological diversity. This includes promoting sustainable agriculture, responsible forestry practices, water conservation measures, and minimising habitat fragmentation.
- **Conservation of Keystone Species:** Protecting and conserving keystone species is vital for maintaining ecological diversity. Keystone species play a significant role in shaping ecosystems and disproportionately impact the structure and functioning of their habitats. By conserving keystone species, we ensure the preservation of ecological interactions and processes.
- **Connectivity and Corridors:** Creating ecological corridors and maintaining connectivity between protected areas and habitats facilitate the movement of species, gene flow, and the exchange of environmental processes. This helps sustain ecological diversity by reducing populations' isolation and supporting ecosystems' natural functioning.
- **Integrated Ecosystem Management:** Adopting an integrated ecosystem management approach considers the complex interactions and interdependencies within ecosystems. It recognises the need to balance conservation goals with sustainable use of natural resources, promoting human well-being while safeguarding ecological diversity.
- **Education and Awareness:** Increasing public awareness and promoting education about the importance of ecological diversity is crucial for its conservation. Encouraging individuals, communities,

and stakeholders to participate in conservation efforts actively can foster a sense of responsibility and stewardship towards ecosystems.

- **Collaboration and Policy Support:** Collaboration among governments, conservation organisations, researchers, and local communities is vital for effective ecological diversity conservation. Supporting policies, legislation, and international agreements prioritising ecosystem conservation and restoration is essential for achieving long-term environmental sustainability.

Conserving ecological diversity is essential for ecosystems' resilience and functioning, ecosystem services provision, and the well-being of both wildlife and human populations. By protecting and preserving the variety of ecosystems on Earth, we can ensure the continued existence of diverse species, maintain ecosystem functions, and sustain our planet's biodiversity.

**Fish conservation involves** various measures to protect fish populations, their habitats, and the overall health of aquatic ecosystems. Key measures commonly employed for fish conservation:

- **Fishing Regulations:** Implementing and enforcing fishing regulations is crucial for sustainable fish conservation. These regulations may include restrictions on fishing gear, size limits, bag limits, closed seasons, and catch quotas. They help prevent overfishing, minimise bycatch, and ensure that fishing practices are environmentally responsible.
- **Protected Areas and Habitat Conservation:** Establishing protected areas, such as marine reserves, sanctuaries, and fishery management zones, helps conserve fish populations and their habitats. These areas provide safe havens for fish to breed, grow, and migrate, contributing to aquatic ecosystems' overall health and diversity.
- **Stock Assessment and Monitoring:** Monitoring and assessing fish populations are essential to understanding their status, abundance, and health. Scientific surveys, data collection, and stock assessments help determine the population dynamics, fishing impacts, and conservation needs for effective management.
- **Habitat Restoration and Enhancement:** Restoring degraded habitats and enhancing fish habitats are critical for conservation. Activities may include reforestation, wetland restoration, removal of barriers to fish migration, creation of artificial reefs, and protection of spawning grounds. These measures improve habitat quality, provide shelter and food sources, and support fish populations.

- **Pollution Control:** Reducing pollution and improving water quality are vital for fish conservation. Measures to control pollution from industrial, agricultural, and domestic sources help protect fish habitats, prevent water contamination, and minimise the negative impacts on fish populations.
- **Invasive Species Management:** Invasive species can harm native fish populations and ecosystems. Implementing strategies to prevent the introduction and spread of invasive species, along with adequate control and eradication measures, is crucial for fish conservation.
- **Research and Conservation Education:** Researching fish ecology, behaviour, and population dynamics provides valuable insights for conservation efforts. Conservation education and awareness programs are vital in promoting responsible fishing practices, sustainable resource use, and public engagement in fish conservation.
- **8. International Cooperation:** Collaborating with other countries and engaging in international agreements and organisations facilitate the conservation of migratory fish species that cross national boundaries. International cooperation helps address the challenges of overfishing, habitat destruction, and illegal fishing on a global scale.

## In-situ conservation

*In-situ* conservation refers to preserving species and ecosystems in their natural habitats. It involves protecting and managing the areas where species naturally occur to ensure long-term survival. In-situ conservation is crucial for maintaining biodiversity, preserving natural ecosystems, and supporting the ecological processes that sustain life on Earth. Some key aspects and strategies of in-situ conservation:

- **Protected Areas:** Establishing protected areas, such as national parks, wildlife sanctuaries, biosphere reserves, and nature reserves, is a primary approach to in-situ conservation. These areas provide legal protection and management measures to safeguard biodiversity, habitats, and ecological processes.
- **Habitat Restoration and Management:** Restoring and managing degraded habitats within protected areas and other natural landscapes is essential for in-situ conservation. This includes reforestation, wetland restoration, grassland management, and controlling invasive species to enhance habitat quality and support native species.

- **Species Monitoring and Research:** Regular monitoring and scientific research help assess the status of species, populations, and ecosystems. Monitoring efforts focus on species distribution, abundance, and ecological interactions, providing valuable data for informed conservation decisions and adaptive management.
- **Community-Based Conservation:** Engaging local communities in conservation efforts is vital for the success of *in-situ* conservation. Involving communities in decision-making, sustainable resource use practices, and conservation initiatives fosters a sense of ownership, promotes conservation awareness, and ensures the long-term sustainability of conservation efforts.
- **Conservation of Key Ecological Processes:** *In-situ* conservation aims to protect individual species and the ecological processes that sustain biodiversity. This includes maintaining natural disturbance regimes (e.g., fire), promoting connectivity between habitats, preserving keystone species, and supporting critical ecological interactions such as pollination and seed dispersal.
- **Landscape-Level Conservation:** The larger landscape context is essential for effective in-situ conservation. Connecting protected areas through ecological corridors and promoting landscape-level conservation planning help maintain ecological connectivity, facilitate species movement, and enhance ecosystem resilience.
- **Indigenous and Traditional Knowledge:** Recognising and integrating indigenous and traditional knowledge systems into in-situ conservation efforts can provide valuable insights into local biodiversity, ecological processes, and sustainable resource management practices. Indigenous communities often have deep environmental knowledge and practices contributing to the conservation and sustainable use of natural resources.

## Fish Sanctuary

A fish sanctuary, or a fish refuge or reserve, is a designated area within a water body established to protect fish populations, their habitats, and the overall aquatic ecosystem. Fish sanctuaries are created to enhance fish conservation, promote sustainable fishing practices, and support the recovery of fish populations. Fish sanctuaries vary in size and design, ranging from small protected areas within a river or lake to more enormous marine reserves covering extensive coastal or offshore locations. Establishing and managing fish sanctuaries involve collaboration between government authorities, local communities, conservation organisations, and fisheries stakeholders. Their effectiveness relies on properly enforcing fishing regulations, community participation, and continuous monitoring to ensure the sustained conservation of fish populations and their habitats.

Some key features and purposes of fish sanctuaries:

- **Protection of Fish and Habitats:** Fish sanctuaries are established to provide a haven for fish species, particularly during critical life stages such as spawning or breeding. These areas offer protection from fishing pressure, habitat destruction, and other human activities that can adversely affect fish populations.
- **Conservation of Biodiversity:** Fish sanctuaries contribute to the conservation of fish biodiversity by safeguarding various species and their habitats. They can serve as important refuges for endangered or threatened fish species, helping to prevent their decline and potential extinction.
- **Enhancement of Fish Stocks:** Fish sanctuaries can support the enhancement of fish stocks by allowing fish populations to grow and replenish within protected areas. The protected environment provides favourable conditions for fish reproduction, growth, and survival, which can lead to increased fish abundance in the surrounding regions through natural dispersal.
- **Research and Monitoring:** Fish sanctuaries often serve as living laboratories for scientific research and monitoring programs. Researchers can study fish behaviour, population dynamics, and the ecological interactions within the refuge, generating valuable data for fisheries management and conservation efforts.
- **Sustainable Fishing Practices:** Fish sanctuaries can act as demonstration sites for promoting sustainable fishing practices and responsible resource management. By highlighting the benefits of protecting fish populations and habitats, they raise awareness among local communities and fishers, encouraging the adoption of sustainable fishing methods outside the sanctuary boundaries.
- **Education and Ecotourism:** Some fish sanctuaries also serve educational and ecotourism purposes. They provide opportunities for people to learn about fish ecology, conservation, and maintaining healthy aquatic ecosystems. Ecotourism activities, such as snorkelling

or guided tours, can generate income for local communities and promote conservation awareness.

### Ex-situ conservation

*Ex-situ* conservation refers to the preservation of species outside their natural habitats. It involves the removal of individuals or populations from their native environments and their placement in controlled settings, such as zoos, botanical gardens, seed banks, or captive breeding programs. *Ex-situ* conservation is employed when *in-situ* conservation measures alone may not be sufficient to ensure the survival and recovery of threatened species. Some key aspects and methods of *ex-situ* conservation:

- **Captive Breeding Programs:** Captive breeding involves breeding and rearing endangered species in controlled environments to increase population numbers. Captive breeding programs often focus on species that face significant threats in the wild, have low population numbers, or suffer from reproductive challenges. These programs can help maintain genetic diversity, provide a source for future reintroductions, and serve as an insurance population against extinction.
- **Seed Banks and Gene Banks:** Seed banks or gene banks preserve the genetic material of plants in the form of seeds, spores, or tissue cultures. These repositories store various plant species' genetic diversity for future use in restoration, research, and breeding programs. Seed banks are critical in conserving plant species, especially rare, endemic, or facing threats.
- **Botanical Gardens and Arboreta:** Botanical gardens and arboreta maintain living collections of plants, including endangered and rare species. These institutions provide a controlled environment for conservation, research, education, and public engagement. They can contribute to the preservation of plant diversity, the study of plant biology, and the cultivation of endangered species for potential reintroduction.
- **Zoo and Wildlife Sanctuaries:** Zoo and wildlife sanctuaries serve as *ex-situ* conservation facilities for preserving endangered animal species. They provide care, breeding programs, and education opportunities for visitors. Modern zoos often prioritise species conservation, scientific research, and public awareness, working collaboratively with conservation organisations and participating in breeding programs for threatened species.

- **Cryopreservation and Assisted Reproduction Techniques:** Cryopreservation involves freezing and storing biological materials, such as sperm, eggs, embryos, or tissue samples, at ultra-low temperatures for long-term storage. Assisted reproductive techniques, including artificial insemination and in-vitro fertilisation, can be combined with cryopreserved genetic material to enhance breeding success and genetic diversity in captive populations.
- **Reintroduction and Translocation:** *Ex-situ* conservation efforts often aim to reintroduce individuals or populations into their native habitats. This process involves carefully planning and implementing the release of captive-bred or ex-situ individuals into suitable wild environments. Reintroduction and translocation programs help restore or enhance wild populations of endangered species.

Ex-situ conservation measures are often combined with in-situ conservation strategies to maximise the chances of species survival and recovery. The ultimate goal of ex-situ conservation is to support the long-term viability of species and their ecosystems by preserving genetic diversity, providing a safety net against extinction, and contributing to species recovery efforts in the wild.

## Cryopreservation of Fish Gametes

Cryopreservation of fish gametes, such as sperm and eggs (also known as ova), is used to freeze and store these reproductive cells at ultra-low temperatures for long-term preservation. Cryopreservation allows the storage of genetic material from valuable fish species, including endangered or commercially important ones, for future use in conservation, research, and aquaculture. The primary goal of gamete cryopreservation is to extend the lifespan of gametes for several years while maintaining their fertilising capacity by reducing their metabolic rate through temperature reduction. Although the preservation of fish eggs and embryos through cryopreservation is crucial, considering the role played by mitochondrial DNA, previous attempts have had limited or no success (Hagedorn *et al.*, 1997; Ahmmad *et al.*, 1998; 2002, 2003(a,b), 2004).

On the other hand, cryopreservation of fish spermatozoa offers numerous potential applications, such as conserving endangered fish species by establishing reserves of genetic material for selective breeding, developing desired genotypes through cross-breeding, facilitating easy transportation and time-independent distribution of genetic material between different locations. It also aids in producing fish seed in species with varying rates of maturity between sexes, promotes seed production in species that are sequential hermaphrodites, and helps reduce costs by eliminating the need to maintain male broodstock in hatcheries (Rao, 1989).

## Overview of the process and its applications

- **Collection of Gametes:** Fish gametes are typically collected through various methods depending on the species. For males, sperm can be obtained by stripping the milt from the male's reproductive tract, while for females; eggs can be collected by gently massaging the abdomen or using hormonal induction.
- **Cryoprotectant Addition:** Cryoprotectants, which protect cells from damage during freezing and thawing, are added to the gametes to prevent ice crystal formation and cellular damage. Commonly used cryoprotectants for fish gametes include dimethyl sulfoxide (DMSO), methanol, glycerol, or combinations thereof.
- **Freezing Process:** Using a controlled cooling rate, the prepared gametes are slowly cooled to a very low temperature. Cooling is usually performed in a programmable freezer or liquid nitrogen vapour. The gametes are typically packaged in cryovials or straws to facilitate the freezing process.
- **Storage in Liquid Nitrogen:** Once the gametes have been frozen, they are transferred to liquid nitrogen (-196°C) for long-term storage. Liquid nitrogen provides a stable and ultra-low temperature environment that effectively preserves the genetic material for extended periods.
- **Thawing and Revival:** When the frozen gametes are needed for use, they are thawed by rapidly warming them to the appropriate temperature. The cryoprotectant is diluted, and the revived gametes can be used for fertilisation or artificial insemination.

Jhingran (1982) conducted initial experiments on the temporary preservation of spermatozoa from Indian Major Carps (IMC). It was observed that when the milt was kept in a Ringer-glycerin solution at 28ºC, it maintained its ability to fertilise for 4 hours. Currently, cryopreservation of sperm has proven to be effective for various species of finfish.

## Applications and Benefits of Cryopreservation of Fish Gametes:

- **Conservation of Genetic Diversity:** Cryopreservation allows the preservation of genetic diversity in fish populations, particularly in endangered species. Frozen gametes can be used for artificial reproduction or to introduce genetic diversity into people with limited genetic variation.

- **Aquaculture and Fisheries Management:** Cryopreserved fish gametes can be used in aquaculture and fishery management practices. They can be used for selective breeding, genetic improvement programs, and the production of valuable traits in fish populations.
- **Research and Biotechnology:** Cryopreserved gametes provide a valuable resource for research and biotechnological applications. They can be used in studies of fish reproductive biology, embryonic development, and gene expression, as well as for research on assisted reproduction techniques and genetic engineering.
- **Species Conservation and Reintroduction:** Cryopreservation allows the long-term storage of fish gametes for species conservation efforts. The preserved gametes can be used for population reinforcement, reintroduction programs, or establishing backup populations in case of environmental catastrophes or population declines.

It's important to note that successful cryopreservation of fish gametes can vary among species and requires specific protocols tailored to each species' reproductive biology. Techniques and procedures are continually refined and improved to enhance cryopreserved gametes' viability and success rates in various fish species.

## Important conventions and acts in India and the world to conserve the fish biodiversity

Several necessary acts and conventions are related to biodiversity conservation in India and worldwide. Here are some significant ones:

### In India:

- **Wildlife Protection Act, 1972:** This act provides legal protection to wildlife species and their habitats in India. It regulates hunting, poaching, and wildlife trade and establishes protected areas such as national parks and sanctuaries.
- **Biological Diversity Act 2002** aims to conserve India's biodiversity and promote sustainable use of its components. It establishes the National Biodiversity Authority (NBA) and State Biodiversity Boards to regulate access to genetic resources and ensure equitable sharing of benefits from their use.

- **Forest (Conservation) Act, 1980:** This act governs the diversion of forestland for non-forest purposes. It requires prior approval from the central government for any project or activity involving deforestation or forest destruction.
- **Environmental Protection Act, 1986:** This act provides the framework for environmental protection and pollution control in India. It empowers the central government to protect and improve the environment and sets guidelines for environmental impact assessments.
- **Water (Prevention and Control of Pollution) Act 1974** aims to prevent and control water pollution. It establishes pollution control boards at the central and state levels and lays down regulations for managing water pollution.
- **The Air (Prevention and Control of Pollution) Act of 1981,** focuses on preventing and controlling air pollution. It empowers the central and state pollution control boards to take measures to prevent, control, and abate air pollution.
- **Coastal Regulation Zone (CRZ) Act 1991:** Coastal Regulation Zone is a regulatory framework established in India to protect and manage coastal areas and their resources. The Coastal Regulation Zone is a specified area along the coastline of India. It includes the land area, water, and airspace extending up a certain distance from the high-tide line. The CRZ Act classifies the Coastal Regulation Zone into four categories, namely CRZ-I, CRZ-II, CRZ-III, and CRZ-IV, based on the ecological sensitivity and development restrictions applicable to each zone.
  - **CRZ-I**: These are ecologically sensitive areas, including national parks, wildlife sanctuaries, mangroves, coral reefs, etc. No new construction is allowed except for projects of national importance.
  - **CRZ-II**: These areas are developed coastal areas with existing settlements. Limited developmental activities are permitted, subject to specific conditions and clearances.
  - **CRZ-III**: These areas are relatively less developed and include rural and urban areas. Certain activities, such as agriculture, fisheries, and infrastructure development, are allowed with necessary clearances.
  - **CRZ-IV**: These areas are designated as the region between the low-tide line and 12 nautical miles seaward. Traditional fishing activities and related infrastructure development are permitted.
- **Marine Fisheries Regulation and Management Act (MFRMA), 1981:** This act regulates and manages marine fisheries in India. It aims to

prevent overfishing, protect fish stocks, and promote sustainable fishing practices. It provides for the declaration of fisheries management zones, licensing of fishing vessels, and regulation of fishing gear.

- **National Fisheries Policy, 2020:** The National Fisheries Policy outlines the government's vision and strategies for the sustainable and inclusive development of fisheries in India. It focuses on promoting responsible fishing practices, enhancing aquaculture, protecting fishery resources, and improving the livelihoods of fishing communities.
- **Inland Fisheries and Aquaculture Improvement Act 2021** aims to regulate and promote inland fisheries and aquaculture activities. It provides for the sustainable development of inland fisheries, protection of fish habitats, and regulation of fish seed production and marketing.
- **Indian Fisheries Act, 1897:** This act addresses offences related to fisheries, such as using prohibited fishing gear, unauthorised fishing, and pollution of water bodies. It prescribes penalties for these offences.
- **Integrated Coastal Zone Management (ICZM) Guidelines:** The ICZM Guidelines provide a framework for the sustainable management of coastal areas, including fisheries and aquaculture. They promote integrated planning, conservation of biodiversity, and community participation in coastal resource management.
- **Indian Fisheries and Aquaculture Standards (IFAS):** IFAS are voluntary standards developed by the Marine Products Export Development Authority (MPEDA) to ensure quality and safety in fisheries and aquaculture practices. They cover various aspects of fish and fishery products' production, processing, and trade.
- **Coastal Aquaculture Authority Act of 2005** establishes the Coastal Aquaculture Authority (CAA) to regulate coastal aquaculture activities and prevent the adverse impact of shrimp and fish farming on the coastal environment.

**World wide**

- **Convention on Biological Diversity (CBD):** This international treaty, ratified by 196 countries, including India, aims to conserve biodiversity, promote sustainable use of its components, and ensure fair sharing of benefits from genetic resources. It was adopted in 1992 during the Earth Summit in Rio de Janeiro.

- **CITES (Convention on International Trade in Endangered Species of Wild Fauna and Flora):** CITES is an international agreement regulating endangered species' trade. It aims to prevent the unsustainable exploitation of wildlife and ensure their survival. India is a signatory to CITES.
- **Ramsar Convention on Wetlands:** The Ramsar Convention is an international treaty that promotes wetlands' conservation and wise use. It recognises the ecological importance of wetlands and their role in biodiversity conservation. Several wetlands in India are designated as Ramsar sites.
- **Convention on Migratory Species (CMS):** Also known as the Bonn Convention, CMS is an international treaty that aims to conserve migratory species and their habitats. It provides a framework for international cooperation to address threats faced by migratory species. India is a party to CMS.
- **FAO Code of Conduct for Responsible Fisheries (CCRF) 1995:** sets out international principles and standards of behaviour to ensure effective conservation, management, and development of marine and freshwater living aquatic resources. 12 articles work
  - **Article 1:** Nature and Scope of the Code
  - **Article 2:** Objectives of the Code
  - **Article 3:** Relationship with other international instruments
  - **Article 4:** Implementation monitoring and updating
  - **Article 5:** Special requirements of developing countries
  - **Article 6:** General Principles
  - **Article 7:** Fisheries Management
  - **Article 8:** Fishing operations
  - **Article 9:** Aquaculture Development
  - **Article 10:** Integration of fisheries into coastal area management
  - **Article 11:** Post-harvest practices and trade
  - **Article 12:** Fisheries research

These acts and conventions play a crucial role in biodiversity conservation in India and globally by providing legal frameworks, promoting international cooperation, and raising awareness about preserving our natural heritage.

# IUCN

The IUCN Convention refers to the International Union for Conservation of Nature (IUCN) and its various conventions and agreements to promote and protect nature and biodiversity. The IUCN is a global organisation composed of governmental agencies, non-governmental organisations, and individual experts dedicated to the conservation and sustainable use of natural resources. It was established in 1948 and has become a leading authority on environmental issues.

## Major fisheries related institute in India

- **Indian Council of Agricultural Research(ICAR**) New Delhi. The Indian Council of Agricultural Research (ICAR) is the apex body for coordinating, guiding, and managing research and education in agriculture in the entire country under the aegis of DARE, Ministry of Agriculture and Farmers Welfare. The ICAR-AU System of India has 74 Agricultural Universities comprising 63 State Agricultural, Veterinary, Horticultural and Fisheries Universities (SAUs), 4 ICAR-DUs, viz. IARI, IVRI, NDRI and CIFE, 3 Central Agricultural Universities (CAU, Imphal, Dr. RPCAU, Pusa and RLB CAU, Jhansi), 4 Central Universities (CUs) having Faculty of Agriculture (BHU, AMU, Viswa Bharati and Nagaland University).
- **ICAR-Central Institute of Fisheries Education** (CIFE): Established in 1961, located at Mumbai CIFE is India's premier fisheries research and training institute. It offers undergraduate, postgraduate, and doctoral programs in fisheries science and conducts research in various aspects of fisheries and aquaculture.
- **ICAR-Central Marine Fisheries Research Institute** (CMFRI): Founded in 1947, located at Kochi, Kerala CMFRI is one of India's oldest marine fisheries research institutes. It focuses on research related to marine fisheries, mariculture, fishery resource management, and conservation.
- **ICAR-Central Institute of Freshwater Aquaculture** (CIFA): Established in 1987, located at Bhubaneswar, Odisha CIFA is dedicated to research and development in freshwater aquaculture. It studies fish breeding, genetics, nutrition, and farming technologies for various freshwater fish species.
- **ICAR-Central Institute of Brackishwater Aquaculture** (CIBA): Founded in 1987, located at Chennai CIBA specialises in research and development related to brackishwater aquaculture systems. It focuses

on technologies for farming brackishwater-adapted fish, shrimp, and other species.

- **ICAR-Central Inland Fisheries Research Institute** (CIFRI): CIFRI was established in 1947, located at Barrackpore West Bengal and focuses on inland fisheries research and development. It studies fish biology, stock assessment, fish genetics, and fisheries management in freshwater bodies.
- **ICAR-Central Institute of Fisheries Technology** (CIFT): Founded in 1954, located at Kochi, Kerala CIFT is dedicated to developing post-harvest technology and value addition in fisheries and aquaculture. It focuses on improving fish processing, preservation, and product development.
- **ICAR-National Bureau of Fish Genetic Resources** (NBFGR) - was established in December 1983 in a rented premises at Allahabad under the aegis of Indian Council of Agricultural Research to undertake research related to the conservation of fish germplasm resources of the country. The Institute's vision is assessment and conservation of fish genetic resources for intellectual property protection, sustainable utilization and posterity..
- **ICAR-Central Institute of Fisheries Nautical and Engineering Training** (CIFNET) - Established in 1963. Located at Kochi, Kerala Its plays a pivotal role in shaping the blue revolution of the country by creating technical manpower for marine operation including fishing, navigation and marine engineering and related shore establishments.
- **ICAR-Central Institute of Coastal Engineering for Fishery** (CICEF), located at Bangalore, Karnataka formerly known as Pre-Investment Survey of Fishing Harbours (PISFH) is a premier institution of the Government of India established in January 1968 with its jurisdiction covering the entire Indian coastline for conducting engineering and economic investigations and preparing the Techno-Economic Feasibility Reports (TEFRs) for the fishery harbour (FH) projects. Prior to 1968, when the maritime States/UTs in the country lacked the experience and expertise required in conducting the required investigations and surveys, PISFH was set up as a Central Project under the Government of India, Ministry of Food and Agriculture, Community Development (Department of Agriculture) in association with Food and Agriculture Organisation (FAO) of United Nations (UN).

- **National Institute of Oceanography** (NIO) - Established in 1966, located at Dona Paula, Goa (conducts research related to marine sciences, including fisheries and oceanography).
- **ICAR- CSSRI:** Central Soil Salinity Research Institute, Karnal, Haryana
- **ICAR- CIARI:** Central Island Agricultural Research Institute, Port Blair
- **ICAR-CIPHET:** Central Institute on Post Harvest Engineering and Technology, Ludhiana, Punjab
- **ICAR-IISWC:** Indian Institute of Soil and Water Conservation, Dehradun, Uttarakhand
- **ICAR- IVRI:** Indian Veterinary Research Institute, Izzatnagar, Uttar Pradesh
- **ICAR- MGIFRI:** Mahatma Gandhi Integrated Farming Research Institute, Motihari, Bihar
- **ICAR-IIFSR:** Indian Institute of Farming Systems Research, Modipuram, Meerut, U. P.
- **ICAR-IIWM:** Indian Institute of Water Management, Bhubneshwar, Odisha
- **ICAR- CIWA:** Central Institute of Women in Agriculture, Bhubneshwar, Odisha
- **ICAR-**National Research Centre on Integrated Farming, Piprakothi, Motihari, Bihar
- **FSI**- Fishery Survey of India, Mumbai
- **MPEDA-** Marine Products Export Development Authority, Kochi
- **RGCA-** Rajiv Gandhi Centre for Aquaculture, Chennai
- **NIFPHATT-** National Institute of Fisheries Post Harvest Technology and Training, Kochi
- **INCOIS-** Indian National Centre for Ocean Information Services, Hyderabad
- **CICEF-** Central Institute of Coastal Engineering for Fishery (CICEF), Bangalore

## References

Ahammad, M. M., Bhattacharyya, D., & Jana, B. B. (1998). Effect of different concentrations of cryoprotectant and extender on the hatching of Indian major carp embryos (*Labeo rohita, Catla catla,* and *Cirrhinus mrigala*) stored at low temperature. Cryobiology, 37(4), 318-324.

Ahammad, M. M., Bhattacharyya, D., & Jana, B. B. (2002). The hatching of common carp (*Cyprinus carpio* L.) embryos in response to exposure to different concentrations of cryoprotectant at low temperatures. *Cryobiology,* 44(2), 114-121.

Ahammad, M. M., Bhattacharyya, D., & Jana, B. B. (2003). Hatching of common carp (*Cyprinus carpio* L.) embryos stored at 4 and −2° C in different concentrations of methanol and sucrose. *Theriogenology,* 60(8), 1409-1422.

Ahammad, M. M., Bhattacharyya, D., & Jana, B. B. (2003). Stage-dependent hatching responses of rohu (*Labeo rohita*) embryos to different concentrations of cryoprotectants and temperatures. *Cryobiology,* 46(1), 2-16.

Ahammad, M. M., Bhattacharyya, D., Banerjee, R. D., & Bandyopadhyay, P. K. (2004). Toxicity of Protectants to Embryos of Silver Carp (*Hypophthalmichthys molitrix*) After Cold Storage at Different Storage Time Periods. *Cell Preservation Technology,* 2(3), 227-233.

Berkes, F. (2007). Community-based conservation in a globalized world. *Proceedings of the National academy of sciences,* 104(39), 15188-15193.

DeGeorges, P. A., & Reilly, B. K. (2009). The realities of community based natural resource management and biodiversity conservation in Sub-Saharan Africa. *Sustainability,* 1(3), 734-788.

Hagedorn, M., Hsu, E., Kleinhans, F. W., & Wildt, D. E. (1997). New approaches for studying the permeability of fish embryos: toward successful cryopreservation. *Cryobiology,* 34(4), 335-347.

Jhingran V G. 1982. Fish and Fisheries of India. Revised edition. *Hindustan Publishing Corporation* (India), G-U-B, Jawahar Nagar, New Delhi 666.

Rao, B., Soufir, J. C., Martin, M., & David, G. (1989). Lipid peroxidation in human spermatozoa as relatd to midpiece abnormalities and motility. *Gamete research,* 24(2), 127-134.

Sarker, F. C., Rahman, M. K., Sadat, M. A., Shahriar, A., & Nowsad Alam, A. K. M. (2022). Haor-Based Floodplain-Rich Freshwater Ichthyofauna in Sylhet Division, Bangladesh: Species Availability, Diversity, and Conservation Perspectives. *Conservation,* 2(4), 639-661.

# Index